The Waterside Ape

The Waterside Ape
An Alternative Account of Human Evolution

Peter Rhys-Evans

CRC Press
Taylor & Francis Group
Boca Raton London New York

CRC Press is an imprint of the
Taylor & Francis Group, an **informa** business

CRC Press
Taylor & Francis Group
6000 Broken Sound Parkway NW, Suite 300
Boca Raton, FL 33487-2742

Printed and bound by CPI Group (UK) Ltd, Croydon, CR0 4YY on acid-free paper

International Standard Book Number-13: 978-0-3671-4548-4 (Paperback)
978-0-3671-4551-4 (Hardback)

Library of Congress Cataloging-in-Publication Data

Names: Rhys Evans, P. H., author.
Title: The waterside ape : an alternative account of human evolution / Peter Rhys-Evans.
Description: Boca Raton : Taylor & Francis, 2019. | Includes bibliographical references and index.
Identifiers: LCCN 2018052422| ISBN 9780367145484 (paperback : alk. paper) | ISBN 9780367145514 (hardback : alk. paper) | ISBN 9780429032271 (ebook) | ISBN 9780429629419 (pdf) | ISBN 9780429627774 (epub) | ISBN 9780429626135 (mobi/kindle)
Subjects: LCSH: Human evolution. | Human beings--Origin.
Classification: LCC GN281.4 .R48 2019 | DDC 599.93/8--dc23
LC record available at https://lccn.loc.gov/2018052422

Visit the Taylor & Francis Web site at
http://www.taylorandfrancis.com

and the CRC Press Web site at
http://www.crcpress.com

This book is dedicated
to
Fran, Olivia, Sophie and James
and to the memory
of
Elaine Morgan

Contents

Foreword   xi
Acknowledgements   xv
Author   xvii
Introduction   xix
Historical Timeline   xxiii

1 Theories of Human Evolution   **1**
   Humanity's Place in Evolution   3
   The Savannah Theory   6
   The Waterside Theory   7
   The Cradle of Humankind and the Influence of Climatic Change   9

2 The Aquatic Debate   **13**
   What Is the Evidence?   20

3 Our Genetic Heritage   **23**
   Mendel and His Experiments   25
   Conditions of Existence   27
   The Late Miocene Drought   28
   The Great Rift Valley   30

4 Our Early Ancestors   **35**
   Early Bipedal Hominins   35
   Australopithecus   36
   Human Populations in the Pleistocene Era   38
       Homo habilis   39
       Homo erectus   39
       Homo heidelbergensis   40
       Homo floresiensis   41
       Homo denisova   41

5 The Neanderthals and Their Demise   **43**

6   The Waterside Ape: Why Are We So Different?   **51**
Bipedalism   53
Nakedness   53
Subcutaneous Fat   54
Thermoregulation   55
Big Brains   58

7   The Naked Ape   **61**
Structure and Function of the Skin   61
Apocrine Glands   63
Eccrine Glands   64
Wrinkly Fingers   65
Subcutaneous Fat   67
Evolutionary Evidence for Changes in Mammalian Skin   68

8   Why We Lost Our Coats: The Early Hominin Tailors   **73**
Desmond Morris and Hominin's 'Christening Ceremony'   75
Skin Colour, Ultraviolet Irradiation and Cutaneous Cancer   77
The Significance of Hominid's 'Christening Ceremony'   81

9   Evolutionary Adaptations in the Human Skull and Sinuses   **83**
Evolution and Comparative Anatomy   89
Maxillary Sinus   89
Frontal and Sphenoidal Sinuses   90
Ethmoid Sinuses   90
Theories of Paranasal Sinus Function   92

10   Human Skull Buoyancy and the Diving Reflex   **95**
The Diving Reflex, Nitric Oxide and the Paranasal Sinuses   96
The Nasal Valve   98
The External Nose   99
The Functional Role of the Nose and Sinuses in a Waterside
    Habitat   100

11   Surfer's Ear   **103**
Ear Canal Bone Abnormalities   105
Developmental Embryology of the Human External Ear Canal   105
Evolutionary Adaptations to the Hearing Mechanism in Aquatic
    and Semiaquatic Mammals   108
External Ear Canal Exostoses in Modern Populations   110
External Ear Exostoses in Archaeological Populations   111
Ear Exostoses as a Vital Fossil 'Missing Link'   112

12 Evolution of the Human Brain **115**
Brain Structure 116
Evolutionary Importance of Lipids 118

13 Food for Thought and the Cognitive Revolution **123**
Fire, Food and Domestication 126
Coastal Migration and Worldwide Dispersal 127
The Cognitive Revolution 128
Genetic Factors Contributing to Modern Brain Evolution 131

14 The Human Larynx and Evolution of Voice **135**
The Upper Airway and Digestive Tracts 137
Diving and Breath-Holding in Hominids 141
Delayed Descent of the Human Larynx 143
Evolutionary Changes in the Brain for Speech and Language 143

15 Obstetric and Neonatal Considerations **147**
The Obstetric Dilemma 148
Fontanelles and Skull Sutures 148
Brain Size and Perinatal Considerations 150
Lanugo and Fat Babies 151
Vernix Caseosa 152
Aquatic Adaptations of Human Babies 153

16 Marine Adaptations in the Human Kidney **155**
The Great Rift Valley 155
Hominid Kidney Adaptation to a Waterside Aquatic Habitat 158

17 Scars of Evolution **161**
Mammalian Evolution 161
Hominin Evolution 162
Adaptations to Bipedalism 163
Lumbar Disc and Sciatic Problems 165
Vertigo, Neck Pain and Why Giraffes Don't Get Dizzy 167
Blood Pressure and Salt Regulation 168
Hernias, Haemorrhoids and Prolapses 170
Varicose Veins 171
The Medical Consequences of Our Bipedal Heritage 172

18 We Are What We Eat   **175**
   Early Hominins and the Crucial Role of Marine and
       Lacustrine Foods   177
   The Cognitive Revolution   179
   Agricultural Origins and the Neolithic Revolution   181
   The Global Food Crisis   183

19 An Incredible Journey   **185**
   The Origin of Speech and Language   187
   The Challenge of Mental Ill Health   189
   Health, Population Growth, Social Inequality and Poverty   190
   An Incredible Journey   191

Glossary   193
References   199
Index   219

Foreword

Back in 1972, when I was in my early twenties, my mother wrote a book. Her name was Elaine Morgan and the book was called *The Descent of Woman*. The book, which became an international best seller, was inspired by Sir Alister Hardy's observation that human skin, with its underlying layer of fat and humans' streamlined body shape, might have evolved as an adaptation to an aquatic lifestyle rather than to life on the savannah.

For Elaine, this was one of those 'Aha'! moments, like Newton's apple falling from the tree or Archimedes' bath overflowing the tub. The more she looked into it, the more sense it made. As she researched the differences between apes and humans, she found that most of our unique characteristics, which don't occur in other primates, are common in aquatic species. In addition to the hard evidence of human physiology and anatomy, Elaine also applied her incisive intelligence and practical common sense to the subject of human origins and demonstrated, to the general satisfaction of her readers, that the savannah hypothesis could not account for these aquatic features.

Some academics however were unconvinced. One prominent anthropologist declared, to Elaine's great amusement, that early humans could never have spent time in the water, because while they were bathing on a summer holiday, small fish had nibbled his penis. Others said that the theory could not be valid because Elaine was a playwright and screenwriter. This seemed a rather curious objection since the savannah or killer ape theory had originally been popularised by Robert Ardrey, who was himself a playwright and screenwriter, and that whole theory was predicated on the shape of a single hominid skull discovered by Raymond Dart in 1924. One common criticism was that the aquatic theory was pseudoscience, because it did not present any falsifiable hypotheses – this being the generally accepted definition of pseudoscience – but that was to change.

Over the years, Elaine wrote several other books on the subject as more and more evidence came to light, and the aquatic ape theory (AAT) slowly but surely began to move into the mainstream.

The Discovery Channel filmed a documentary about it featuring Desmond Morris, and *The Descent of Woman* became a set book in American universities. In 1987, there was a major conference on the subject at Valkenburg in the Netherlands, followed by conferences and symposia in Southampton,

Sun City, Oslo, Ghent; and, in 2013, in London, chaired by Peter Rhys-Evans of the Royal Marsden Hospital, the author of this book.

Elaine's online TED (Technology, Entertainment and Design) Talk has now been viewed over a million times, and *The Descent of Woman* has been translated into more than 25 languages. In recent years, Sir David Attenborough has presented two BBC Radio 4 series – *The Scars of Evolution*, and, in 2016, *The Waterside Ape*, based on my mother's work. Through that second series, I became acquainted with Peter Rhys-Evans.

I had made a couple of brief contributions to the programme, so I was interested to see what the public reaction would be to this account of the latest developments in the field. Online comments from the public were encouraging, making for a lively debate with good, rational arguments on both sides of the 700-plus posts on one Web site.

The academic response, though, was disappointing. Mark Maslin, a geography professor and climatologist, together with TV pundit Alice Roberts, published an article headed 'Sorry David Attenborough, we didn't evolve from 'aquatic apes' – here's why' (punctuation per the original work).

I wondered why the article was so superficial, but then I learned that neither of the authors had actually listened to the programmes, and they were quite rightly called out on it by a dozen professors from research institutions all over the world. Maslin and Roberts' arguments, for what they are worth, included a number of claims about the 'overwhelming evidence' against the theory arising from the 'huge advances (that) have been made in the study of human evolution'. For some reason, though, they omitted to mention any of the evidence or the advances.

They did, however, come up with several untestable hypotheses, such as, 'adaptations to heat loss better explain our pattern of body hair'. That is, we lost our fur to keep warm?

Another unsupported, untestable hypothesis was, 'Voluntary breath control is more likely to be related to speech than to diving', though I don't personally know anyone who holds their breath while speaking, and howler monkeys manage to vocalise quite effectively without doing so.

Yet another – 'Sexual selection may also explain our body fat distribution'. Now I have a noticeable accumulation of fat around my midriff. I'm not aware of any women finding that distribution particularly attractive, but what do I know?

Elsewhere, though, the same writer has stated, 'Distribution of fat in humans is much better explained as a dietary adaptation'. I suppose if you come up with enough random, untestable hypotheses then one of them could be right.

The little 'evidence' that is presented is also not very persuasive: 'Compared with other animals, we are not actually that good at swimming'.

Really? Well we can swim faster than sea otters, dive as deep as the harbour porpoise and hold our breath for longer, swim further than water voles both on the surface and underwater and dive from a greater height than any other mammal. Not bad for a savannah ape.

They save their most compelling evidence against the aquatic theory till last: 'our fingers become prune-like after a long bath'. In fact, any fisherman will tell you that the one thing they are most able to grip better with wrinkled fingers is ... fish – an unlikely adaptation on the savannah.

One thing on which I am in wholehearted agreement with Maslin and Roberts is that 'we must always build our hypotheses on, and test them against, the hard evidence: the fossils, comparative anatomy and physiology, and genetics'. In this book you will read about some of the many testable hypotheses that the AAT has generated over the years.

One hypothesis based on comparative anatomy and tested against fossil evidence was this: Since frequent immersion in cold water causes exostoses to form in the ear canals of modern humans (surfer's ear), if our distant ancestors foraged for food underwater, then fossil skulls should also exhibit such exostoses, which has now been demonstrated in fossil hominid skulls 1–2 million years old.

Another fact, based on physiology and tested via comparative anatomy, concerned an organic compound called squalene. Squalene is present exclusively in the skin of water mammals, from whales to otters. If humans are aquatic apes, then their skin should also produce squalene, which it does.

Again, human babies are the only primates that are covered in vernix when they are born. If this is an aquatic adaptation, then other marine mammals should exhibit the same feature, which some, like the semiaquatic seal, certainly do.

In the meantime, the savannah theory has only ever generated one falsifiable hypothesis – that if early hominids lived on the savannah, then plant and animal remains in the vicinity of their excavated skeletons should be from the savannah. The presence of liana vines and marine and aquatic fossils have failed to support the hypothesis.

In the absence of testable hypotheses, the savannah theory must actually be nothing more than pseudoscience, and disproved pseudoscience at that. There is by now such a wealth of literature of all kinds on the AAT that no book could possibly aim to encompass all the accumulated evidence. The idea has taken on a life of its own. Not only are there countless academic papers, books and articles on the subject, it has also caught the public imagination to such an extent that writers of all sorts have published short stories, children's books and even pop songs about 'sea apes'.

One article, in the May 2018 edition of *Scuba Diving* magazine, includes the experimental results of yet another hypothesis. This was formulated to answer a question that occurred to me during the BBC Radio 4 series 'The Waterside Ape' hosted by Sir David Attenborough consisted of 2 programmes on consecutive evenings. My question was this: If early hominins evolved during the Miocene drought, whether on the savannah or on the sea coast, what did they do for fresh water? Could it be that we are able to absorb adequate quantities of fresh water from the sea through our sweat glands without drinking at all?

Alister Hardy first presented his aquatic hypothesis to members of the British Sub-Aqua Club in 1960, so it is fitting that they should be the first to read the initial findings on this question, which could confirm the AAT beyond any reasonable doubt.

In 2004, Elaine said that she thought that within 10 years AAT would be 'over the cusp' and in the mainstream of evolutionary anthropology. That was certainly the case in terms of public support and, with this pivotal book, Peter Rhys-Evans has pushed the weight of evidence well beyond the tipping point.

Peter Rhys-Evans is a consultant ENT (Ear, Nose and Throat) surgeon, so it is no surprise that his knowledge of human anatomy with regard to some of the organs most relevant to aquatic adaptation is both comprehensive and detailed. Yet he writes in a style that makes his expertise accessible to the non-technical reader as well as satisfying the most stringent academic requirements.

I thought I had pretty much kept abreast of research relevant to AAT, but I found more than a dozen astonishing facts that were completely new to me in this invaluable reference work. I also found a number of new, as yet untested hypotheses, including an exciting and very credible mechanism for the evolution of the descended larynx. These will, I'm sure, result in many more revelations in the future. There has never been a better time to be an evolutionary anthropologist.

Gareth Morgan

Acknowledgements

I would like to pay tribute to Sir Alister Hardy and to all those who were enthralled by the books written by Elaine Morgan and continue to believe in the Aquatic Ape Hypothesis proposed by her. I am also very grateful to Elaine's son Gareth for agreeing to write the Foreword for this book.

There are many distinguished scientists and academics who have made significant new discoveries and contributed to the 'Aquatic' debate over recent decades. In particular I would like to thank Michael Crawford, Marc Verhaegen and Mario Vaneechoutte for their contributions and support. Others include Erik Abrahamsson, Nick Ashton, Geoff Bailey, Dame Valerie Beral, Bernard Bichakjian, Bruce Bradley, Tom Brenna, Leigh Broadhurst, Javier Caraveo-Patino, Brunetto Chiarelli, Stephen Cunnane, Andreas Fahlman, Clive Finlayson, Nic Flemming, Robert Foley, Anna Gislen, Naama Goren-Inbar, Cedric Hassall, Donald Johanson, Orio Johansson, Sir David King, Algis Kuliukas, Martha Lahr, Jeffrey Laitman, John Langdon, Dirk Meijers, Stephen Munro, Michel Odent, Stephen Oppenheimer, Michael Petraglia, Pierre-Francois Puech, Erika Schagatay, Dennis Stanford, Kathlyn Stewart, Sir Kenneth Stuart and Philip Tobias.

I am grateful to Michaella Cameron and to Margaret Ijegbai and her team at the Royal Marsden Hospital for their help in organising the Evolution Conference in 2013 and to Jimmy Mulville, Tony Matharu and the late Sir Adrian Swire, without whose generous sponsorship the meeting could never have taken place.

I would also like to thank the team at CRC Press/Taylor & Francis and all those who have helped in the production of this book. In particular I would like to thank Miranda Bromage to whom I sent my original manuscript and to Chuck Crumly who believed in the work. It has been a pleasure to work with them and Jennifer Blaise, Linda Leggio and Melissa Dalton. Jason Candlin has done many of the illustrations and I am grateful to him and to Julia Gould who kindly suggested a quotation. My special thanks also go to Veronique Baxter and Sir Michael Morpurgo for their valuable support and advice.

Above all, I am indebted to Sir David Attenborough for his support and encouragement to all those who have tried to unravel new facts about early human evolution. It was he who organised a symposium on 'The Aquatic Ape' in 1992 with Elaine Morgan, helped with the evolution conference in 2013 and produced the BBC Radio 4 programme in 2016. His tireless work has been inspirational in discovering new and fascinating insights into the beautiful but dangerous natural world on Planet Earth, where evolutionary adaptation to the changing environment is crucial for survival and procreation.

Finally, I would like to thank my wife Fran and our three wonderful children who have patiently put up with my indulgence in apes and 'fishy' matters, which have proved to be a fascinating distraction from my 'day job' as a surgeon over the past 30 years.

Author

Peter Rhys-Evans was born in Kent in 1948 and was educated at Ampleforth College in Yorkshire. After qualifying as a doctor from St Bartholomew's Hospital in 1971, he spent some time as a lecturer in anatomy at Bristol University Medical School. He then specialized in ear, nose and throat (ENT) and plastic and reconstructive surgery before spending a year (1980–1981) studying for a postgraduate degree in head and neck cancer surgery at the University of Paris, while he was a resident ENT surgeon at the Gustave Roussy Institute, Paris.

In 1981, he was appointed as Consultant ENT Surgeon at Birmingham University Queen Elizabeth Hospital, and in 1986 he was invited to take on the post of Chief of ENT/Head and Neck Surgery at the Royal Marsden Hospital in London. He gave up his National Health Service (NHS) work after 35 years in 2016, but continues in private ENT practice. As an early pioneer in new techniques of laser and reconstructive surgery since the 1980s, Dr. Rhys-Evans and his team have developed many innovative techniques over the years.

With over 250 scientific publications, including 5 books, he has been an active contributor to the specialty. His award-winning textbook *Principles and Practice of Head and Neck Surgery and Oncology*, published in 2003 with a second edition in 2009, won a prestigious award from the University of London for the 'Best International Publication in ENT Surgery' during the preceding five years. His valued reputation in his speciality is reflected in his membership of national and international societies and committees, and in presenting over 320 major lectures in 26 different countries. He recently delivered the prestigious 2017 Royal College of Surgeons *Arris & Gale* lecture on a new controversial theory of early human evolution, a research interest since the 1980s.

He is the founder and executive chairman of Oracle Cancer Trust, which is now the largest charity in the United Kingdom, raising funds for head and neck cancer research. Since 2001, he has been responsible for raising over £6 million and, with his colleagues, for establishing a very active research programme. Apart from supporting groundbreaking PhD research projects for improving survival outcomes in head and neck cancer, Oracle's other main objective is to fund research into restoring quality of life and reducing side effects and complications following treatment of patients.

Dr. Rhys-Evans lives in Sussex with his wife, Fran, and their three children, together with a menagerie of dogs, chickens, ducks, pigs and other animals.

Introduction

*To raise new questions, new possibilities,
to regard old problems from a new angle,
requires creative imagination and marks real
advance in science.*
Albert Einstein (1879–1952)

As a recently appointed consultant ear, nose and throat (ENT) surgeon with a particular research interest in the nose and sinuses, I was asked in 1985 to write a chapter in the then standard textbook of ENT surgery on the subject of 'Anatomy and Physiology of the Nose and Para-Nasal Sinuses'. From my days at medical school, I recalled the words of the Renaissance physician Jean François Fernel in 1548: 'Anatomy is to Physiology as Geography is to History – it describes the Theatre of Events'.

In my endeavour to write something new and interesting about the subject, I looked at the comparative skull anatomy of humans and other primate species. I was puzzled about the major differences in our anatomy and physiology compared with our primate cousins, the gorillas and chimpanzees, not only in the head and neck area, but all of the other unique characteristics which are not found in any other terrestrial mammal. I found it difficult to reconcile these distinctive human features with the traditional belief in the savannah theory of evolution.

By chance, someone mentioned to me an interesting new book on evolution by Elaine Morgan entitled *The Aquatic Ape*. I was intrigued by the proposed marine scenario of hominid evolution and realised that many of the unanswered questions I had about some of the anatomical and physiological differences between ourselves and other primates and terrestrial mammals were much more logically explained by this theory, and that the development of early humans was influenced by a period of aquatic adaptation.

My chapter on the nose and sinuses was duly written, but I was interested to examine in more detail these exceptional evolutionary differences in the head and neck area from a medical point of view. I wanted to see if I could find a more logical explanation rather than just consider evolution from fossil evidence. I discussed some ideas with Elaine Morgan, and in 1992, I subsequently wrote an article in an ENT journal which included some aquatic-related theories about our unique anatomical and physical differences.

One of these related to a curious condition in which bony swellings appear to grow in the deep part of the ear canal. ENT surgeons have known about these so-called exostoses for over a century, and we see them quite often in our clinic patients, but no one has been able to explain why they occur. They can be seen quite easily when the ear is inspected and are usually small, but they can grow to quite a size and may even close off over 80% of the ear canal so that the eardrum is difficult to see. The unusual aspect of these bony protrusions is the fact that they are only seen in people who swim frequently and in surfers, or rather, in people who are associated with frequent water immersion of the ear canal. Consequently, many patients are somewhat surprised when, having looked in their ears which might be quite asymptomatic, I might say: 'Oh, I see that you are a keen swimmer!' followed by the inevitable response: 'How on earth did you know that?'

In the general population, exostoses occur in about 2–6% of people, but in regular surfers or swimming instructors, they can be present in 70–90%. As a result, the condition is now colloquially known as surfer's ear. The size and incidence of these swellings seem to be directly related to the frequency and length of time the ear is exposed to water, especially cold water.

Much has been written about the possible cause of these bony swellings, but several aspects still remain unclear. The fact that the condition is almost always bilateral is fairly predictable since both ears are immersed in water, but why do they only grow in swimmers? Why do they only grow in the deep part of the ear canal and not elsewhere in the body which is exposed to water? Also, why do they always grow at two or three constant sites near the eardrum. And, from an evolutionary point of view, what is or was the purpose and function of these rather incongruous bony protrusions?

This strange anomaly is just one of a number of anatomical and physical features that we doctors see in our patients, as well as various ailments and medical conditions which are unique to humans and not seen in any other higher primates nor in any other terrestrial mammals. These differences are difficult to reconcile with the traditional theory of evolution, which suggests that our early hominin ancestors, having split from apes and chimpanzees about 6–7 million years ago, evolved on the savannah as bipedal hunter-gatherers.

Mammals, after all, had evolved over a period of 50 million years to what seemed an ideal shape and form – a head at one end and a tail at the other, with a horizontal spine between the two, supporting the internal structures, and a leg at each corner providing optimal stability. Many of the medical conditions seem to be due to the fact that humans, for some reason, decided to stand on two legs rather than four and that our spine is vertical, rather than horizontal, unlike any other land mammal. The additional pressure in the lower parts of our bodies would explain conditions such as hernias, prolapses, haemorrhoids, spinal disc problem and varicose veins. But why would one single branch of the primate family undergo evolutionary changes which would make us so different from our ape cousins and all other terrestrial mammals and, as a result, cause such additional medical problems?

The two other African apes in our ancestral family, the gorilla and the chimpanzee, like other higher primates such as the bonobo, have remained unchanged in appearance and behaviour for over 20 million years, but why did one branch of the ape family evolve a dramatic change in its physical characteristics and in its mode of life to eventually become the hairless, bipedal, big brained early hominin humans? It was not, as Elliot Smith in the 1920s and others believed, that *Homo sapiens* was the 'pinnacle of evolution' and that humans were the destiny of the evolutionary process. The emergence of modern humans from our lowly origins as quadruped apes 6–7 million years ago has been due to the one simple motivating factor common to all evolutionary changes – the quest for survival. In her book, *The Scars of Evolution*, Elaine Morgan explains, '*Man* is no more an evolutionary pinnacle than a tree is, or a termite or an octopus. His emergence was no more inevitable than that of any other species'.[1]

Despite a great deal of criticism and ridicule from the press and mainstream anthropologists about this controversial aquatic or waterside theory of evolution, it is gratifying that in the last 25 years many of our predictions have been verified and the waterside ape theory is now beginning to be accepted by some. We are grateful that two of the greatest visionaries in human anthropology and natural science, Professor Phillip Tobias, the South African anthropologist, and Sir David Attenborough have realised the validity of this controversial theory of early hominin evolution and have contributed their support for further research and interest in this field.

There are many excellent books on human evolution which concentrate on essential fossil and archaeological evidence to demonstrate the ancestral lineage of hominins, but very few of them mention this alternative aquatic theory at all. This book, however, is written from a medical point of view in an attempt to explain the unique differences in comparative anatomy and physiology between humans, primates and other terrestrial mammals in a logical fashion. It contains many new concepts based on scientific evidence published over the past 20 years which provide rational explanations and also important fossil evidence that our early hominin ancestors spent a prolonged period of evolution adapting to a semiaquatic habitat.

Little did our ENT colleagues realize that an explanation of the cause and formation of ear exostoses might help in understanding the evolutionary process of early hominins from our ancestral ape family. I hope that *The Waterside Ape* will convince readers of the validity of the aquatic or waterside ape theory that early humans evolved from our arboreal ape ancestors with unique characteristics, not seen in any other terrestrial mammals, which allowed us to become the dominant global mammalian species *Homo sapiens* ('wise man').

It will, I hope, at least provoke some constructive scientific debate on early hominin evolution and help persuade critics to produce some valid counter-arguments to explain these differences, rather than simply dismiss the theory and disregard the scientific evidence.

It may be difficult, however, for some to accept the fact that there is now substantial evidence that we have evolved from our ancestral ape family as a semiaquatic mammal and not as a terrestrial savannah ape, as is generally believed. To my mind, the waterside ape theory provides the only logical reason for our many unique primate characteristics and helps to explain how our early hominin ancestors struggled, survived and successfully evolved to eventually become the humans we are today.

Historical Timeline

YEARS BEFORE PRESENT

4.54 billion: Earth formed, approximately one-third the age of the universe. The early atmosphere contained virtually no oxygen. Much of the earth was molten with extreme volcanic activity. Over time the earth cooled, causing the formation of a solid crust allowing water to condense on its surface.

3.8 billion: Beginning of life on earth with single cell protozoa.

3.2–2.4 billion: Photosynthetic organisms began to enrich the atmosphere with oxygen.

600 million: Atmospheric oxygen levels rose sufficiently for air-breathing lifeforms to emerge.

580 million: The Cambrian fossil record from China shows intracellular membrane synthesis, possibly utilizing oxygen with lipids and proteins, allowing more complex multicellular organisms to develop. Docosahexaeonic acid (DHA) was one of the key components for new photoreceptors and nerve signalling in marine life evolution and has remained a vital substance for neural development in most species.

541 million: Complex lifeforms, including vertebrates, and the first fishes began to dominate the oceans around 530 million years ago and gradually life expanded onto land with formation of plants, animals, insects, reptiles and fungi. Birds, the descendants of dinosaurs, and more recently mammals emerged.

500 million: First brain structure appeared in worms (hindbrain).

250 million: Appearance of new paleo-mammalian region of the brain.

66 million: A 10 km asteroid struck earth causing a huge gas cloud that caused a loss of sunlight, inhibiting photosynthesis. Seventy-five percent of all life, including non-avian dinosaurs, became extinct.

60 million: Mammalian diversity increased, some of which (e.g. cetacea) took to the oceans to eventually evolve into whales; others, like primates, took to the trees.

6–7 million: First evidence of genetic split between chimpanzees and part bipedal/part arboreal hominid.

4.5 million: Appearance of Ardepithecines who were bipedal but also probably part arboreal.

3.5 million: Bipedal Australopithecines appear, still with small brains.

2.5 million: Evidence of *H. habilis* and use of tools, with larger brains (700 cc).

2 million: *H. erectus* found in Java following the first 'Out of Africa' exodus, with enlargement of the cerebral cortex (880 cc).

1.3 million: Likely start use of fire, mainly for heat and protection.

600,000: *H. heidelbergensis* appears with brain capacity around 1,200 cc.

500,000: Neanderthals evolve in Europe and Asia with large brains.

300,000: Regular daily use of fire for cooking.

200,000: *H. sapiens* evolved in east Africa.

100,000: Anatomically modern humans (AMH) appeared, and *H. sapiens* migrated to Eurasia in the second 'Out of Africa' exodus.

70,000: The 'Cognitive Revolution' began with larger communities, trading, creative art and speech.

45,000: *H. sapiens* reached Australia.

30,000: Extinction of Neanderthals.

16,000: Humans settled in America.

13,000: Extinction of *H. floresiensis* in Indonesia, leaving *H. sapiens* as the only surviving human species.

12,000: Onset of the 'Agricultural Revolution' with domestication of plants and animals, with more permanent 'waterside' settlements, notably in the Levant and around the 'Fertile Crescent'.

Theories of Human Evolution

1

Scientists have become the bearers of the torch of discovery in our quest for knowledge.
Stephen Hawking (1942–2018)

Charles Darwin is now rightly recognised as one of the most important and influential scientists in history. However, in the middle of the nineteenth century, his radical theory of evolution, based on evidence he had gathered years before on his voyages on the *Beagle* to the Galapagos Islands, was something that he found extremely awkward to share with others and difficult to publish.

In Victorian England, the church had immense power and influence over people's lives, and it took Darwin 20 years to pluck up the courage to publish *The Origin of Species* in 1859. He was a shy and humble person who was plagued by ill health and knew that his theories would be severely criticised by the church and the establishment. It was only because he was concerned that another young naturalist, Alfred Wallace might pre-empt his ideas that he finally agreed to publish. He realised that he was not a great speaker and was not well enough to expound his theories in person; instead, he relied on other naturalists, notably Joseph Hooker and Thomas Huxley, to champion his work.

In the introductory paragraph of *The Origin of Species*, Darwin hesitatingly writes: 'until recently the great majority of naturalists believed that species were immutable productions and had been separately created. Some few naturalists, on the other hand, have believed that species undergo modification, and that the existing forms of life are the descendants by true generation of pre-existing forms'.[1] It was only in the last chapter of his book that he makes any comment or reference about human evolution, with the immortal scientific phrase that *light will be thrown on the origin of man and his history.*

The publication of *Origin of Species* was followed in 1871 with Darwin's book *The Descent of Man* which expounded his theory of evolution of humans and apes from a common ancestor.[2] There is little dispute today about the validity of this theory, although there is much argument about the manner in which the remarkable differences between the higher apes and humans were evolved (Figure 1.1). In Darwin's time, however, there were only a few odd

MR. BERGH TO THE RESCUE.

THE DEFRAUDED GORILLA. "That *Man* wants to claim my Pedigree. He says he is one of my Descendants."

Mr. BERGH. "Now, Mr. DARWIN, how could you insult him so?"

FIGURE 1.1 Illustrated *London News* 1859. (Mr. Bergh was President of The Royal Society for the Prevention of Cruelty to Animals.) (Courtesy of *Harper's Weekly.*)

bones of a pre-historic Neanderthal man from Belgium, Germany and others from Gibraltar, and his concern was 'whether man, like every other species, is descended from some pre-existing form'.[2]

More recently, Elaine Morgan writes: 'Considering the very close genetic relationship that has been established by comparison of biochemical properties of blood proteins, DNA structure and immunological responses, the differences between a man and a chimpanzee are more astonishing than the resemblances'.[3] These include structural differences in the skeleton, the muscles, the skin and the brain; differences in posture associated with a unique method of locomotion; differences in social organization; and finally the acquisition of speech and tool using, which together with the dramatic increase in intellectual ability, has led scientists to name their own species, *Homo sapiens* – wise man.

There is little doubt that the three main higher primate species – the gorilla, the chimpanzee and pre-hominid ape – evolved from the common ancestral African ape. But what were the circumstances which dictated such a divergent evolutionary path for hominins during which they acquired unique adaptations such as walking on two legs, a bigger brain, loss of body hair, subcutaneous fat, an excess of sweat and sebaceous glands, and changes in sexual and social habits? Hominins also demonstrated modifications to the upper airway and digestive tract, which were not seen in any of the other apes. We are no more special than any other species in evolutionary terms, although we have uniquely evolved the intellectual capacity to analyse our past history with the benefit of a modern understanding of evolution and genetics.

HUMANITY'S PLACE IN EVOLUTION

There are many gaps in the fossil record and our understanding of the evolutionary pathway of early humans (hominin species), but piecing the scant evidence together with the help of genetic analysis gives us a broad outline of how *humans* evolved from the first bipedal *Australopithecines* about 3–4 million years ago (Figure 1.2). There are several distinct milestones where new characteristics or features are apparent, either in the fossil record of ancestral hominin skulls or in early humans' associated habitats or diet. There is also evidence of changes in their lifestyle and community interactions, quite apart from technical and intellectual

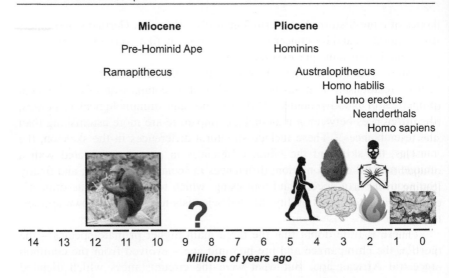

FIGURE 1.2   The evolutionary 'Black Hole'. (Courtesy of the author.)

advances. These important developments include bipedalism, use of tools, bigger brains, use of fire, ceremonial burials, and eventually the significant artistic and intellectual advances associated with the 'cognitive' revolution around 70,000 years ago.

It is difficult to explain many of these unique characteristics in terms of traditionally held beliefs of human evolution from the 'Savannah Ape', which assumes that ancestral hominin species evolved by changing from an arboreal habitat in the trees to a new lifestyle on the open grassland. Some other vital factor or circumstances must have played a role in motivating these unique adaptations, which are clearly more advantageous in setting humans totally apart from their primate cousins on a path of evolution leading ultimately to the emergence of *Homo sapiens.*

The crucial period of time during which these remarkable divergent evolutionary changes started to take place in the higher primates was in the late Miocene/early Pliocene epoch, which commenced about 9 million years ago and lasted roughly 5 million years. Fossil remains of ape-like pre-hominid primates during the preceding Miocene period are abundant both in Africa and Asia, and one bearing early changes in jaw structure thought to be possibly precedent to the hominin ape was Ramapithecus, discovered by G. E. Lewis in India in 1930 and later by Leakey in Africa. During this period, large populations of ape-like creatures flourished in temperate forest expanses widely distributed throughout Asia, Africa and Europe.

What happened next over the next few million years during the so-called evolutionary 'Black Hole' is uncertain, however, because there is little fossil evidence during this period. This may partly be explained by the fact that much of this period was during the Ice Ages and sea water levels were much lower because of the extensive frozen ice caps; indeed, many of the archaeological sites are now under water.

The iconic image of the quadruped monkey on the left with various images of ape/human-like creatures gradually becoming more upright and less hairy, leading to the recognizable human image on the right (Figure 1.3), is mere conjecture because we do not know precisely what and when these changes took place.

One of the main uncertainties in understanding evolution for much of the hundred years after Darwin was the important question about which came first, a bigger brain or bipedalism. Did our early ancestors acquire a larger brain and improved intelligence, which prompted them to think that standing upright would allow them to free their upper limbs and hands to be able to carry weapons and develop improved manual dexterity? Or did they acquire bipedalism out of survival necessity and then realise that they were able to use their hands and forelimbs for purposes other than quadruped locomotion?

Fossil evidence has shown that the majority of ape-like species, including the chimpanzee, gorilla, bonobo and orangutan, evolved with few changes during the Pliocene period and to the present time. However, remains discovered in the Olduvai Gorge in East Africa in the 1970s from about 3.5 million years

FIGURE 1.3 Iconic Image of Ape/Man transition. (Courtesy of Science Photo Library/Science Source.)

ago indicate that a dramatic evolutionary step had occurred in one particular branch of the ape family. *Australopithecus* ('southern ape'), as he was called, was different from all other apes in that he walked upright on two legs instead of four. His skull, however, was still the same size as that of a chimpanzee, confirming that bipedalism came well before a bigger brain.

The vital questions about bipedalism, brain evolution or any other aspect relating to differences between apes and humans were not really considered at the time of Darwin. Because the apes were tree dwellers and humans lived on *terra firma*, it was assumed that humans had simply dropped down from the trees onto the savannah and had stood upright because he 'could see further' and became a 'hunter-gatherer'. In retrospect, this seems to have been a rather irrational assumption of *post hoc ergo propter hoc* ('after this and therefore because of this'). Since there was little archaeological evidence to suggest otherwise and the mechanisms of evolutionary changes were poorly understood, there was never any consideration of an intermediate phase or other explanation for the significant differences that were to distinguish hominins from other primates.

Ever since then scientists have accepted the 'savannah' theory and have dismissed, criticized or ignored new ideas such as those proposed by Hardy[4] and Morgan.[3] There have been many excellent books on human evolution, but none of them have offered a plausible explanation for the unique characteristics seen in humans compared with other primates. It is largely thanks to Elaine Morgan, Professor Philip Tobias[5] and Sir David Attenborough[6] that the debate has been kept alive. Others, some looking from a medical rather than an archaeological perspective, have had the basic conviction that evolution may have followed a slightly different path from the savannah scenario.[7–10]

THE SAVANNAH THEORY

This traditionally held 'savannah' theory of evolution proposed in the nineteenth century and supported by Dart in 1924 has held sway for most of the twentieth century. This postulates that gradual evolution of ancestral humans from forest arboreal apes occurred because of climatic and behavioural changes. Loss of their forest habitat as a result of the late Miocene drought with corresponding extension of grassy plains or savannahs resulted in movement of pre-hominoid apes from the tree habitat on to the plains. No longer able to depend on lush vegetation in the forest, the savannah apes evolved an omnivorous diet scavenging for small game to satisfy their needs, later evolving into hunters. According to this theory, the one crucial development of bipedalism in the savannah ape evolved because of the advantage of standing upright on two legs and being able to see further over the plains and high grass in search of prey and to avoid predators. Later, the advantage of having two hands free to carry weapons enabled humans' development as hunters in pursuit of game.[11]

But did we stand upright in order to hunt and kill big game on the plains of South Africa, as Dart and screen writer, Robert Ardrey have written?: 'Man has emerged from the anthropoid background for one reason only, because he was a killer'.[11] Or did we find a much less demanding environment where the food sat obligingly there on the river bed and sea floor for us to collect at our leisure without the need to chase after wild game and antelopes?[6]

One of the major weaknesses of the savannah theory is the disproportionate extent of major differences between humans and their ape cousins in physical and physiological characteristics, for example, bigger brains and bipedalism. If this suggested move on to the savannah and standing upright was so advantageous for one branch of the ape family, why have we not seen any similar or parallel developments in some of our primate cousins or in any other savannah mammal? None of the other apes have acquired even one of our unique characteristics. Other ape species have remained as gorillas and chimpanzees with few morphological or physiological changes over 20 million years up to the present time. Others, such as the patas and vervet monkeys and the baboons, also moved onto the savannah from the trees but, despite this change of habitat, they have evolved with little change or physical modification.

THE WATERSIDE THEORY

The first person to question the traditional 'savannah' theory of evolution was Max Westenhofer, a German biologist and pathologist who first suggested that water might have played an important role in the evolution of humans. His book *Der Eigenweg des Menschen* (*The Unique Road to Man*), published in 1942, included ideas suggesting that water adaptation had played a significant role in human development.[12] However, he disputed Darwin's basic concept of humans' evolution from apes and thought that humans had a more immediate aquatic past and had only recently returned to land.

In April 1960, however, an article appeared in the *New Scientist*[4] written by Sir Alister Hardy, an eminent marine biologist, endorsing the aquatic theory of evolution which he thought seemed to explain many of the dramatic changes that occurred during humans' development from the arboreal ape. He noticed that many of the unique characteristics seen in humans and not in any other terrestrial mammal were often present in aquatic and semiaquatic mammals. He proposed that, as a result of the changing environment at the onset of the Miocene drought, with reduction of the forest habitat and vegetation, certain anthropoids, ancestral to humans, were driven by competition from life in the trees to assume a new habitat on the shores of inland waters, estuaries and the seashore. In this environment they became adept at hunting for abundant food, shellfish and sea urchins in the shallow waters. The co-incidental inundation of the Afar region in East Africa forming the Great Rift Valley would have provided an ideal environment for these changes.

Initially wading on all fours, these 'aquatic' apes gradually assumed a more upright posture because of the necessity to keep their heads and airway out of the water. As a result, the pelvis gradually rotated to a more vertical alignment, giving the body a more streamlined shape. They gradually were able to extend their territory into deeper water, and greater versatility on land and in the water enabled them to escape more easily from predators, both on land and in the water. Later came the ability to swim and dive and, because of the additional buoyancy, stability on two legs would have been much easier to acquire in the water environment than on the savannah, as exemplified by the only other primate to have ventured into the sea, the proboscis monkey (Figure 1.4).

The proboscis monkey lives in the mangrove swamps in the coastal waters of Borneo, and, among other features, the male of the species is characterized

FIGURE 1.4   The proboscis monkey. (Courtesy of Don Mammoser/Shutterstock.)

by his enormous nose. Although retaining a mainly terrestrial quadruped gait similar to other apes living in trees, he can adopt a bipedal mode when wading in the shallows in order to keep his head above water. In certain instances, he has even been noted to use this mode of movement for walking on dry land.

The compelling argument proposed by Hardy that ancestral humans spent a period of aquatic adaptation is not unique in evolutionary history. On the contrary, this process is well recognized, and several species of birds, reptiles and mammals are known to have abandoned their terrestrial existence to become adapted and modified to an aquatic life. An early example is a member of the dinosaur family (ichthyosaur), which took to the water, evolving flippers instead of legs before becoming extinct. Among mammals, members of the cetacean family (whales, dolphins and porpoises) successfully adapted to an aquatic existence, as did some of the hoofed mammals related to the elephant (the sea cows and manatee). Others include aquatic birds (penguins), aquatic carnivores (sea lion, seal, otter), aquatic rodents (beaver, water vole), aquatic reptiles (crocodile, sea snake) and aquatic insectivores (water shrew, desman).

There is therefore no reason to suppose that a similar transformation did not occur among the primate order. The waterside theory postulates that one such species of ape did embark on this course and that adaptation to this new semiaquatic environment resulted in the gradual emergence of a bipedal ape, Australopithecus, ancestral to *Homo erectus* and *Homo sapiens*.

THE CRADLE OF HUMANKIND AND THE INFLUENCE OF CLIMATIC CHANGE

It is universally accepted that the evolution of early hominins took place in the region of the Great Rift Valley in East Africa. Until recently, virtually all authors have accepted that early hominins evolved from our primate quadruped forest-based ancestors as terrestrial mammals who adapted to a life on the savannah. There have been several general concepts proposed about how early hominins might have evolved, but these are all based on the 'savannah' theory.

In 1985, Vrba first suggested that global climate change was a cause of African mammalian evolution[13] which led to the development of the *turnover pulse hypothesis*. More recently, an alternative *variability selection hypothesis* has been proposed by Potts[14,15] suggesting that 'a long-term trend with increasingly complex intersection of orbitally-forced changes in isolation and earth-intrinsic feedback mechanisms results in extreme, inconsistent environmental variability selecting for behavioural and morphological mechanisms that enhance adaptive variability'.[16] For some, it might be difficult to interpret or understand this theory of evolution.

There is evidence of extreme climatic and environmental variability during the Plio-Pleistocene period, and marine records of African climate variability describe a shift toward more arid conditions about 2.8 million years ago. This would have had a major effect on the vegetation of East Africa and human evolution.[15,17] For example, it is acknowledged that hominin speciation (formation of new species) events and changes in brain size seem to be statistically linked to the occurrence of transient deep-water lakes in the Rift Valley and that the emergence of *Homo erectus*, with their big brain, around 2 million years before present (mybp), was co-incidental with the period of maximal lake coverage.[18]

Undoubtedly, these important climatic and environmental changes would have had a profound effect on the general habitat in the region of the Great Rift Valley and would have influenced periods of migration depending on the presence or absence of lakes and fresh water and the aridity of the vegetation. However, changes in the climate at this time would not have been instrumental in causing the fundamental and radical physiological or anatomical characteristic differences between apes and early hominins such as bipedalism, loss of hair, subcutaneous fat, differences in thermoregulation and so on, which had already evolved some 3 million years previously towards the end of the Miocene era.

Indeed, our ancestors, their primate cousins and other terrestrial mammals would all have been exposed to identical climatic and environmental influences on the ground, whether it was a forest-based or savannah-based habitat. Yet others did not develop any similar adaptations or any of the characteristics which were unique to hominins. These evolutionary adaptations therefore need to be explained individually in anatomical and physiological terms rather than in relation to the general climate and other environmental factors.

These major anatomical and physiological changes could only have happened if the actual body of early hominins was exposed to a totally different *milieu extérieur du corps* ('external environment around the body'). If hominins had evolved on the savannah, this would have been no different whether we were in the trees or on the savannah/forest floor. One possible explanation is that this 'immediate' *milieu extérieur* was a habitat that included a substantial element of aquatic foraging, swimming and diving for food, and not a purely terrestrial environment similar to that of our ape cousins, whether on the savannah below or still in the trees.

I believe that it is important to look at each of the unique human characteristics and to try and explain them individually in logical terms.

As Einstein said: 'To raise new questions, new possibilities, to regard old problems from a new angle, requires creative imagination and marks real advance in science'. A great deal of new scientific evidence has been published over the past 30 years which cannot be ignored.

New theories such as the waterside/aquatic ape hypothesis should be looked at and discussed objectively rather than discarded or ignored. In their excellent textbooks, *Evolution – The Human Story* by Alice Roberts[19] and *The Complete World of Human Evolution*[20] by Chris Stringer, there is no mention of the 'aquatic' theory. Apart from the half-centenary publication: *Was Man More Aquatic in the Past? Fifty Years after Alister Hardy*,[10] the only author in the past 10 years to have mentioned the 'aquatic' theory is Dr. Michel Odent, an obstetrician and pioneer of 'birthing pools', who wrote about *The Birth of Homo, the Marine Chimpanzee.*[21]

The Aquatic Debate

2

> *When a new idea is rejected, it is frequently because it flies in the face of an accepted prevailing paradigm, in this case the Savannah Hypothesis.*
> Professor Philip Tobias (1925–2012)

Elaine Morgan was a Welsh writer for television and author of several books. and has been one of the main proponents of the 'aquatic ape' theory of human evolution. Having read the book on evolution *The Naked Ape*, published in 1967, by Desmond Morris about 'man the hunter', she became interested in the inconsistencies in the 'savannah' theory of evolution. In this book Morris makes reference to Hardy's 'aquatic' theory of hominin evolution, but he discounts it because of lack of fossil evidence.[1]

Elaine's first book *The Descent of Woman* (1972) highlighted the apparent discrimination of woman's important role in evolution[2] and was followed in 1985 by her book *The Aquatic Ape Hypothesis*.[3] The scientific community did not take her seriously, however, and criticized her books, treating them with disdain because she was not an academic scientist.

In 1992, Sir David Attenborough organized a symposium on *The Aquatic Ape* at the annual conference of the British Society for the Advancement of Science at which Elaine Morgan and other supporters of the theory were speaking. At the conference, reference was made about the importance of a marine diet for development of the large human brain. The function of the sinuses in lightening the human skull to help buoyancy in an aquatic hominin was also mentioned, as were exostoses found only in the ear canal of frequent swimmers and surfers. A prediction was made that if these small bones were found in early human skulls, it would provide vital fossil evidence of the validity of the 'aquatic ape' theory.[4]

Once again, however, the scientific establishment and the press dismissed the idea. During the week of the conference it was reported in *The Times* with the caption: '*How apemen kept humanity in the mainstream*', accompanied by

FIGURE 2.1   *The Times* front page article on 22 August 1992. (Courtesy of Times Newspapers Limited.)

an early apeman photofit of *'The predecessor of the swimming instructor?'* (Figure 2.1). The following week, also writing in *The Times* in his usual amusing and satirical style, Bernard Levin *'plunges into the Darwinian debate with only his sinuses for support. The missing link – c'est moi!'* (Figure 2.2).

Over the next few years the critics had a field day. In a 1997 critique, anthropologist John Langdon considered the aquatic ape hypothesis (AAH) under the heading of an 'umbrella hypothesis' and argued that it was difficult to disprove such a thing and that it could not serve as a proper scientific hypothesis. Langdon considered the 'savannah' hypothesis to be the 'collective discipline of palaeoanthropolgy'. He observed that some anthropologists had regarded the AAH idea as not worth the trouble of a rebuttal. In addition, the evidence cited by AAH proponents mostly concerned developments in soft tissue anatomy and physiology, whilst palaeoanthropolgy rarely speculated on evolutionary development of anatomy beyond the musculo-skeletal system and brain size as revealed in fossils. His main conclusion was that the AAH was unlikely ever to be disproved on the basis of comparative anatomy and that the one body of data that could potentially disprove it was the fossil record.[5]

One of South Africa's most honoured and decorated scientists, Professor Philip Tobias, was a palaeoanthropologist who was best known for his work on early hominin fossil sites. For much of his career, he was an adherent of

FIGURE 2.2   Bernard Levin's article in *The Times* on 31 August 1992. (Courtesy of Times Newspapers Limited.)

the 'savannah' hypothesis and the belief that our ancestors may have started walking/running on two legs in order to see further across the open plains. In 1985, he wrote:

> I published a chapter called 'The conquest of the savannah and the attaining of erect bipedalism' in which I expressed the old idea: 'The living apes of Africa are to be found exclusively in the wet forest of the middle reaches of the continent. It is likely that ancestral apes, too, were forest-dwelling creatures… The spread of lighter woodland and savannah and the retreat of the margins of the primaeval forests could well have created conditions in which the tendency to uprightness and bipedalism was favoured. The ability to run across the high grass cover of the savannah, perhaps from one woodland-girt stream to another, might have held advantages for those apes which could 'walk tall'. Uprightness gave its possessors a chance to see over the tall grass and to watch out for predatory enemies like the lions and sabre-toothed big cats. Seemingly it was under just such a set of conditions that the Hominidae made their appearance upon the face of the earth. That statement may well be the quintessence of the Savannah Hypothesis (SH) – and I believe it was my last statement in support of it. In 1995, I stated of the SH, 'We were all profoundly and unutterably wrong'.[6]

In 1998, Tobias asked for a renewed interest in the importance of water in human evolution, referring to the work of Elaine Morgan and Marc Verhaegen.

Tobias did not espouse the aquatic ape hypothesis as a primary cause for, for instance, our bipedalism, since there are other theories, but to his knowledge, no other hypothesis could explain those several physiological and biochemical characteristics which seem to ally us to aquatic mammals.

Tobias pointed out that proximity to water was the most important factor in the location of an evolving group like the early hominins. Since humans become quickly dehydrated when denied water in warm, tropical or subtropical climates, our ancestors must have lived in the vicinity of springs, rivers, lakes and freshwater estuaries. Wherever the early members of the human family were evolving, they needed water to drink and to keep cool. Tobias also argued that water must have played a crucial role in the dispersal of humanity across the planet.

> In the face of all this EVIDENCE, old and new, it is time for human evolutionists to open their minds and give fair and objective thought to the role of water in the evolution of mankind. We need a new holistic emphasis on water: first for drinking, secondly as a source of food from aquatic plants and animals and, thirdly, as waterways facilitating - or impeding - the spread of humanity across the globe. Fourthly, we may no longer shy away from the questions posed by those especial features of the human skin, sweat-glands, chemistry of sweat, body temperature control and fluctuations, heat and radiation tolerance and water consumption, which in modern humans appear so different from those of savannah-adapted mammals and so reminiscent, in some cases, of aquatic mammals.[6]

But even such a world-renowned and respected scientist as Phillip Tobias could not break down the wall of silence that existed around any consideration of the aquatic ape theory. His views were also largely ignored.

> 'When a new idea is rejected, it is frequently because it flies in the face of an accepted prevailing paradigm, in this case the Savannah Hypothesis', wrote Prof. Tobias in *Out There*. And he goes on: 'Now, at least, anthropologists should be able to examine this with a more open mind than previously when the thinking of so many was clouded by the Savannah Hypothesis'.[6]

In the light of emerging new scientific evidence, in May 2013 an international conference, *Human Evolution – Past, Present and Future* was organised in London, supported by Sir David Attenborough, Donald Johanson and many other international experts. The arguments for and against the aquatic ape theory were discussed, but by this time the savannah ape theory had largely been discredited by many.

Despite this, critics continued to ignore the increasing scientific evidence. It was all very well for the well-known broadcaster and journalist Bernard

Levin to write his amusing piece in *The Times* two decades before, but one of the senior editors of the scientific journal *Nature*, Henry Gee, wrote during the week of that conference in 2013:

> People think they know about evolution, but the 'aquatic ape' theory isn't science: it's creationism ... an arcane and eldritch item of fashionable nonsense in the shuddering slush-pile of human evolutionary rubbish.[7]

The author did not address any of the arguments put forward at the conference in support of the AAH by many international speakers presenting their evidence.[7]

Anthropologist John Hawks wrote that 'it is fair to categorize the AAH as pseudoscience because of the social factors that inform it, particularly the personality-led nature of the hypothesis and the unscientific approach of its adherents'.[8] Physical anthropologist Eugenie Scott has described the AAH as an instance of 'crank anthropology' akin to other pseudoscientific ideas in anthropology such as alien-human interbreeding and Bigfoot.[9]

Popular support for AAH became an embarrassment to some anthropologists who wanted to explore the effects of water on human evolution without engaging with AAH, which they considered 'emphasizes adaptations to deep water (or at least underwater) conditions'. Foley and Lahr suggest that 'to flirt with anything watery in palaeoanthropolgy can be misinterpreted', but argue, 'there is little doubt that throughout our evolution we have made extensive use of terrestrial habitats adjacent to fresh water, since we are, like many other terrestrial mammals, a heavily water-dependent species'. But they allege that 'under pressure from the mainstream, AAH supporters tended to flee from the core arguments of Hardy and Morgan towards a more generalized emphasis on fishy things'.[10]

In September 2016, Sir David Attenborough hosted another forum on BBC Radio 4 in a two-part series *The Waterside Ape*, which asked some straightforward questions: First, how long have humans and our ancestors been habitual users of aquatic and marine resources? Second, have we adapted physiologically and cognitively to a littoral environment, in which we depended on those aquatic and marine resources? And finally, what evidence from the last 15 years of research has emerged to refute or to illuminate either of these questions?[11]

On September 16, the day after the broadcasts, Alice Roberts, professor of public engagement in science, and Mark Maslin, professor of geography, published a blog on TheConversation.com which was subsequently reprinted in *The Guardian, The Independent and Scientific American*: '*Sorry David Attenborough, we didn't evolve from "aquatic apes" – here's why*'.[12] In the article they made claims which might have been meaningful 40 years ago, but which were no longer of any relevance in 2016. Much of the article was a repeat of what Henry Gee, the *Nature* editor, had written 3 years previously after the international conference in London. Specifically they stated that:

1. There is no fossil or other evidence to support the waterside model, and
2. The waterside model 'makes no falsifiable predictions, therefore it is pseudo-science' (Courtesy of Henry Gee@*Nature*, retweeted by Roberts).

In reply to this article, a number of international academic scientists who supported Sir David Attenborough and the waterside theory wrote:

> Our ancestors may indeed have evolved at the shoreline – and here is why ...[13]
>
> Both of these complaints are examined directly in *The Waterside Ape*. We would direct Roberts and Maslin to the following research that has been published in peer-reviewed journals over the last 15 or so years, all of which was covered in *The Waterside Ape* broadcast, but which they chose to ignore in their article:
>
> 1. Human diving physiology and performance compared with semiaquatic mammals (Schagatay 2014; Schagatay, Fahlman, 2014 – in *Human Evolution*).
> 2. Auditory exostoses in modern humans (surfer's ear) and in fossil skulls going back to 1–2 million years ago provide evidence of frequent swimming in *Homo erectus* and *Homo neanderthalensis* (Rhys-Evans and Cameron, 2014 – in *Human Evolution*).
> 3. Oxygen isotope data showing that early hominids at 2–3 million years ago were habitually in shallow water and depending on wetland sedges and papyrus (Magill et al., 2016 in *PNAS*).
> 4. Predation and preparation of very large catfish in Turkana basin at 2 million years ago (Braun and Archer, 2014 in *Journal of Human Evolution*) and very large carp at the Acheulian site of Gesher Benot Ya'aqov (Alperson-Afil et al., 2009, in *Science*).
> 5. Pachyosteosclerosis, that is, dense and brittle bones in *Homo erectus* suggesting a shallow-diving habit (Verhaegen, Munro, 2011, in *Journal of Comparative Human Biology*).
> 6. Shallow diving for *Euryales ferox* nuts at GBY around 800 thousand years ago (Goren-Inbar et al., 2014, in *InternetArch*).
> 7. Wading and exploitation of large mussels both for food and tools at Trinil, in Java around 500,000 years ago (Joordens, Munro et al., 2015, in *Nature*).
> 8. Dependence on mussels and sea-snails at Pinnacle Point at 164,000 years ago (Marean et al., 2007, in *Nature*).

9. Evolution of the hominid brain requiring iodine, iron, selenium, zinc and other nutrients in addition to docosahexaeonic acid (DHA) (Broadhurst et al., 2002, in *Br J Nutrition*).
10. Vernix caseosa: A falsifiable hypothesis was set up, tested and proven valid that vernix is likely to be an adaptation to entering water soon after being born (Brenna et al., 2016, submitted).

It would be of value to your readers if Roberts and Maslin could explain in what ways the above research fails to support a waterside model of human evolution? Also, in what sense are any of the above authors practitioners of 'pseudo-science'?

Is the message of Roberts and Maslin that the general public should be kept in ignorance of the peer-reviewed research from the last 15 years that points to water-side habitation and adaptation at many points in human evolutionary history? We think the job of engaging the public in science is about encouraging debate and looking at the evidence, rather than attempting to dictate which discussions are scientifically acceptable. That is the job of the peer-reviewed journals.

We would strongly encourage readers to take the out-of-date assertions of Roberts and Maslin with a good tablespoonful of salt, of the iodised variety, and to read the original work to which we refer, then make their own judgements.

We applaud the BBC and Sir David Attenborough for keeping up with the most recent scientific literature and presenting it to the wider public, in contrast to the addressed professors.

Signed,

Erika Schagatay, Professor of Animal Physiology, Mid Sweden University, Sweden

Peter Rhys-Evans, Consultant Otolaryngologist, the Lister Hospital, London, United Kingdom

Kathlyn Stewart, Research Scientist, Canadian Museum of Nature, Ottawa, Canada

Marc Verhaegen, General Physician and Researcher in Human Evolution, Mechelen, Belgium

Mario Vaneechoutte, Professor of Medicine and Bacteriology, University of Ghent, Belgium

Naama Goren-Inbar, Professor of Archaeology, Hebrew University of Jerusalem, Israel

Stephen Munro PhD, Curator at the National Museum of Australia

Algis Kuliukas, PhD, Researcher at the University of Western Australia

Stephen Cunnane, Professor of Medicine, Sherbrooke University, Canada

Tom Brenna, PhD, Professor, Cornell University, United States

Michael Crawford, Visiting Professor, Imperial College, London, United Kingdom[13]

It has been argued that such evolution could not have been possible over a period of 3 to 4 million years, but it is much more likely that these evolutionary changes took place in a relatively short period of time, because our ancestors were adapting to a completely different semiaquatic habitat. The powerful impetus for such major unique changes would not have been the same if they had simply moved from the trees onto the forest floor and grassland where they had already existed with their ape cousins for several million years without any change. Why did one branch of the ape family suddenly embark on such radical changes if it wasn't because they had to survive in a different ecological niche? Increasing desiccation and loss of the forest habitat may have provided that initial environmental stimulus where certain families of apes living near the coastal waters, marshes or riverside were forced to adapt to a more aquatic existence in search of food by wading into the shallow waters.

Under these unusual conditions, evolution of the hairless, bipedal apes described by Hardy and Morgan, may have started to take place. Among them could have been hominoid *Australopithecines*, who ventured south along the Afar Gulf towards the Hadar settlement where fossils were eventually found.[14] Other *Australopithecines* may have migrated to the Koobi Fora and Olduvai Gorge where the remains of *Homo habilis* were found.[15,16] Further evolutionary development continued with emergence of *Homo erectus* about 1 million years before present (mybp), until the final desiccation of the Great Rift Valley about 30,000 years ago.[17,18]

The waterside theory postulates that when these upright *Australopithecines* were on land, various adaptations had evolved as a result of their semiaquatic existence which provided distinct advantages over other ape species on the savannah. Their stable bipedal gait permitted more freedom of the forelimbs when they were on land to enable them to hold and carry objects and weapons, which was distinctly favourable in their emergence as dominant hunters in this new terrestrial environment.

WHAT IS THE EVIDENCE?

Whatever is the truth about evolutionary development of early humans from the arboreal apes, suppositions must be based on established scientific facts, although the final explanation of the sequence of events may not be readily apparent. We only have small pieces of the large jigsaw, but the waterside theory does seem to provide a more logical and consistent answer to many of the enigmatic inconsistencies between higher primates and humans and is frequently supported by parallel development in other species.

Fossils provide the only 'hard evidence' of evolutionary changes and have proved beyond doubt that bipedalism was the first crucial adaptation in early hominin evolution.[19] Recent detailed studies of early skulls and post-cranial remains of *Homo erectus* by Rightmire[20] have also provided useful information about comparative changes in bone structure. Although much of the evidence for Elaine Morgan's 'aquatic theory' relates to soft tissue adaptations such as changes in subcutaneous fat, loss of hair and changes in thermoregulation where specific dating is hard to establish, it is based on sound scientific deductions.[21] Preservation of skeletal remains in a marine environment is extremely unpredictable because of continuous wave erosion, and if the waterside theory is to be believed, it is not surprising that few early hominin fossils dating from this period have been found. Also, since the end of the Ice Age, many of the archaeological sites have been underwater. However, those remains of primitive *Australopithecines* which have been identified have all been in the vicinity of coastal areas or inland waters in the region of the Afar peninsula and Olduvai Gorge, a fact which protagonists of the savannah theory claim is coincidental.

What these critics seem to forget, however, is that there *is* now definite fossil evidence of the waterside theory, which so far has been largely overlooked by anthropologists and scientists but which proves that early humans did venture into the water in search of food, and that, unlike their ape predecessors, they did spend a large proportion of their time swimming and diving. This evidence is crucial to the argument supporting the waterside theory, and with modern computerized tomography (CT) imaging and an increasing number of adequate fossil remains of early anthropoid humans, it will hopefully be possible to obtain more of this evidence, which relates to the presence of ear canal exostoses, better known as the condition surfer's ear.[22,23]

Following his original comments at the time of the international conference in 2013, writing again at the time of Sir David Attenborough's BBC radio broadcast in September 2016, Henry Gee, the senior editor of *Nature* commented[24]:

Why on earth is the BBC stooping once again to give the 'aquatic ape' hypothesis the time of day? It's not a 'theory': it's a fun idea 'supported' by cherry-picked factoids and impossible to falsify. It's not science.

Fortunately, however, the author then reveals his source of information.[24] He goes on to say:

If none of this convinces you, consider this thought experiment, which comes from that well-known source of knowledge and wisdom – playground humour.
Q: Why do elephants paint the soles of their feet yellow?
A: So they can hide upside down in bowls of custard.

All I can say to those who may be reluctant to consider recent scientific evidence is that the proof is in the pudding. Their hominin ancestors fought and struggled to survive over several million years to evolve eventually into the dominant, bipedal, intelligent hominin primate species *Homo sapiens* that we are today. Thanks to their aquatic and marine diet, they evolved a large brain and far superior intelligence than any other species.

It would be helpful if evolutionists could embrace the recent scientific evidence supporting the waterside theory rather than ignore it, and engage in meaningful discussion and debate rather than repeat outdated comments that are no longer relevant. After all, hominin evolution is not only concerned with fossil evidence of bones and the size of the cranial cavity, adaptations have also involved important changes in the anatomy, physiology and biochemical processes in the body over millions of years. There are many senior anthropologists, scientists, physicians, naturalists and academics who have devoted years of research into understanding different aspects and theories of evolution.

I firmly believe that the scientific evidence for the aquatic/waterside theory, which I will discuss in subsequent chapters, provides a logical explanation for the various unique adaptations in hominin evolution. Thanks to their aquatic bipedal gait, humans can stand upright and be counted and, as a result of their aquatic-adapted throat and descended larynx, they can also eat, talk and continue to argue and carry out intellectual conversation. If, in the course of time, the waterside theory is finally accepted, I hope that certain critics do not choke when they are obliged to eat their own words in a humble pie, or end up with too much egg on their faces when they realise that the theory is actually true and hide upside down in bowls of custard!

Our Genetic Heritage

3

The Human Genome Project 'is a history book – a narrative of the journey of our species through time. It's a shop manual, with an incredibly detailed blueprint for building every human cell. And it's a transformative textbook of medicine, with insights that will give health care providers immense new powers to treat, prevent and cure disease'.
Francis Collins
Director of National Human Genome
Research Institute (2001)

Over the past 15 years since publication of the Human Genome Project, not a month passes without some new revelation about our genetic heritage or some brilliant work on how to eradicate distressing hereditary diseases. The first complete human genome was published in 2001 after a decade of great endeavour by hundreds of scientists worldwide at a cost of $3 billion. It unravelled the mysteries of the genetic DNA sequences. The 100,000 Genome Project was launched in December 2014 to help clinicians and researchers better understand, and ultimately treat, rare and inherited diseases and common cancers.[1] More than 100 families are being recruited every month, and already there are over 250,000 fully sequenced human genomes with millions of other samples from living and deceased people around the world.

As well as removing a large amount of uncertainty for the families, the results stand to have a major impact on many areas of their lives including future treatment options, social support and family planning. They also have the potential to help many more children with undiagnosed conditions who may be tested for these genetic mutations early on and be offered a diagnosis to help manage their condition most effectively. In the United Kingdom there are plans to consider sequencing genomes from everyone at birth,[1] but already it is now possible for any of us to send off a sample of our saliva for analysis to reveal our ancestral record.

The amount of data that is being produced is phenomenal and, together with archaeological records, it is helping to date much more accurately the sequence of hominid divergence and evolution. Up until recently we have relied on fossil analysis of teeth and bones with relatively crude dating methods, but now, if we are lucky enough to be able to extract DNA from these archaeological remains, we can piece together much more accurate genetic information about early hominins and answer many questions; for instance, what was the relationship between Neanderthals and our ancestral *Homo sapiens*?

Ever since Darwin published his book on *The Origin of Species by Means of Natural Selection* in 1859, the scientific world has been aware that evolutionary changes in a particular species of animal or plant may or may not be beneficial to provide a superior advantage for survival in their particular habitat. The word 'evolution' was not mentioned in Darwin's great work and, although he expounded his theory about the process of natural selection, the phrase 'survival of the fittest' was coined by the philosopher Herbert Spencer to describe the essential process of evolution.

Twelve years later, when Darwin published *The Descent of Man*, there were only a few scant remains of primitive humans from the Neander valley in Germany, a skull from Gibraltar and another from Belgium, and yet Darwin suggested that 'it is somewhat more probable that our early progenitors lived on the African continent than elsewhere'.[2] At that time Darwin was more concerned 'whether man, like every other species, is descended from some pre-existing form ...' and had no idea about how these genetic characteristics and modifications were passed from one generation to the next. It was not until the end of the nineteenth century that the principles of heredity were understood.

We now know that characteristics are passed from generation to generation by replication of genes and that physical differences are brought about by random genetic mutations. Cells in an organism are dividing all the time and mutations occur regularly, producing cells that are slightly different (e.g. different coloured eyes). If this change offers some advantage which results in a better chance of survival and propagation (e.g. blue eyes are more attractive), the mutated change is likely to be continued. Each species will develop a particular niche in their own animal or plant hierarchy, but there are continuous changes in the environment and also rival species may develop a new mutation which offers them a survival advantage. This results in continuous change and evolution, producing species that are better at surviving in a particular habitat. Darwin found many examples of this species adaptation in the finches of the Galapagos Islands during his travels, which inspired his evolutionary theory.

These slight changes in plant and animal characteristics have been recognised and used for thousands of years by farmers, shepherds and herders who have known that by selectively cross-breeding their animals or plants with variations or hybrids that are sturdier or are better suited to a particular environment, they can produce new offspring which are more likely to survive and be more productive.

MENDEL AND HIS EXPERIMENTS

Around the time that Darwin was writing his great work on evolution, a little-known monk in the monastery of St. Thomas in the city of Brunn (now Brno in the Czech Republic) was embarking on some experimental work with plants. The monastery was dedicated to the teaching of science and one of their novices, Gregor Mendel, was sent to Vienna to gain some teaching experience and credentials. However, he failed his examinations and returned to the monastery, where he started doing some experiments on breeding pea plants which he published in 1863.[3] The true significance of his ideas on the principles of heredity was not realised until 1900, long after his death when he was posthumously recognized as the founder of the science of genetics, or how different characteristics or traits are passed from parents to their offspring, which applies to all species in the animal and plant kingdom.

Mendel studied garden peas for three main reasons. First of all, they were readily available and cheap and also came in a variety of shapes, colours and sizes which could easily be distinguished. Second, they could be easily either self-pollinated or cross-pollinated. And third, they could be easily grown in large numbers.

Mendel observed seven different characteristics in the pea plants that were either one of two alternatives (phenotypes). For instance, the peas were either smooth or wrinkly, yellow or green; the plants were either tall or short and the petals were either purple or white. There was never an intermediate variation. Pollination of pure traits over several generations always produced the identical appearance as their parents, but he observed the effects of cross-pollination of two alternative traits (e.g., yellow and green peas) over several generations. He found that crossing a yellow pea with a green pea always produced a yellow pea in the first generation (f1) but in the following generation (f2) after self-pollination there was a consistent 3:1 ratio of yellow to green peas. The same was true for all of the seven different alternative traits.

He concluded that one of the colour traits was inherited from each parent and that one was a stronger 'dominant' yellow factor (Y) and the other was a weaker 'recessive' green factor (g). The offspring of cross-pollination of a yellow (Y) and a green pea (g) (f1 generation) would have the two factors, but the appearance (phenotype) would always be yellow because the dominant

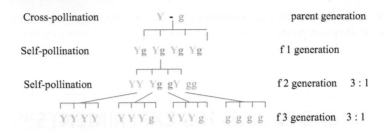

FIGURE 3.1   Mendel's plant experiments. (Courtesy of the author.)

Y factor was present (Figure 3.1). When the offspring (all Yg) were self-pollinated, the alternatives for the offspring (f2) would be YY, Yg, gY and gg; the colour would therefore be yellow in three of them (containing Y) and green in one (where there was gg and no Y). These factors are now known as genes and the genotype is the combination of the genes, either homozygous (where the pair are the same, for example, YY, homozygous dominant, or gg, homozygous recessive) or heterozygous (different, e.g., Yg, or gY). Although the outward appearance (phenotype) of a particular plant may be yellow, the recessive g gene (green) is present and can be passed on to successive generations.

Mendel also showed that these traits or genes are passed on independently from other traits, which explains why in human inheritance, having blue eyes does not increase or decrease the chance of having brown or blond hair. We now know that these different genes regulating different characteristics are located on different chromosomes in the cell and, from analysis of human genomes, we can even predict the colour of someone's eyes.

In the 1940s, it was discovered that DNA was the molecule in the cell that transmitted the genetic information from one generation to the next, but it wasn't until the early 1950s that two scientists, Rosalind Franklin and Maurice Wilkins, studied DNA using X-rays. Franklin produced an X-ray photograph that allowed two other researchers, James Watson and Francis Crick, to work out the three-dimensional structure of DNA, which was found to be a double helix. This gave it the ability to copy itself and to replicate identical cells to pass on from parent to child.[4] Crick and Watson won the Nobel Prize for their discovery and it was a British scientist, Fred Sanger (winner twice of the Nobel Prize), who discovered how to sequence the DNA molecules, letter by letter.

The one driving force in evolution therefore is survival of the species in an ever changing environment and natural selection. We know that 10–15 million years before present (mybp) our ancestors were one of many families of quadruped apes living in the deciduous forests in Africa. However, something must have happened to our branch of the family which didn't happen to the ancestors of apes, gorillas or chimpanzees, or indeed to any other terrestrial mammal.

The unique characteristics of humans that we see today would not have evolved at the same time but gradually over several million years. However, there must have been some initial impetus which started the whole process and the long journey, eventually evolving into *Homo sapiens*. It could have been a change in climate or tectonic activity which altered the habitat in which our ancestors lived, or a series of genetic mutations, or a combination of both. The fight for survival and the threat of extinction in a changed environment, similar to a dramatic change which resulted in the extinction of the dinosaurs, must have been associated with random genetic mutations in our branch of the family and not in any of the other apes, which meant that our ancestors could survive better than their cousins in this altered evolutionary scenario.

Our knowledge about what happened and when has to be based on fossil evidence, which unfortunately is sparse throughout this period. Most of the early hominin fossils have been found in East Africa in the region of the Great Rift Valley and there is general consensus, as Darwin had predicted 150 years ago, that evolution from quadruped ape to bipedal early hominin happened in this region.

CONDITIONS OF EXISTENCE

Darwin considered that, apart from genetic factors influencing evolutionary changes, there were also two other important considerations, namely, the so-called epigenetic forces in evolution. The first of these was 'natural selection' or 'survival of the fittest' (or better adapted to a particular environment). The second of these epigenetic factors was 'conditions of existence', and he thought that 'conditions of existence' were the more powerful. For instance, where identical twins are brought up in two different environments, they will eventually have different characteristics and will behave and respond differently to each other.

These 'conditions of existence' are governed by epigenetic changes in the cellular control system or genome, which is an organism's complete set of DNA, including all of its genes. Each genome contains all of the information needed to build and maintain that organism. In humans, a copy of the entire genome – more than 3 billion DNA base pairs – is contained in all cells that have a nucleus. External events and changes in the environment can control or switch these genes on or off, but they do not change the basic genome in the cells. What they do is to alter which genes are expressed by addition of molecules (e.g., methyl groups) to the basic DNA backbone. This will determine how the organism responds to environmental factors.

THE LATE MIOCENE DROUGHT

So what were the 'conditions of existence' or factors which might have had a major influence on evolutionary changes in hominins and their environment during this crucial period? One possible event was a dramatic climatic change, which was to alter the geological and ecological structure in many of these regions, and in one place in particular, in the Afar region of East Africa. The onset of what was known as the late Miocene/early Pliocene drought between 8 and 5 mybp resulted in loss of the forest habitat for these apes and created a changing environment when adaptation was essential for survival (Figure 3.2).

There is some controversy about the extent of the aridification, but evidence suggests that the climate in South Africa became continuously drier after 8 mybp and that there was maximum C4 (mainly tropical grasses) savannah expansion between 5 and 4.5 mybp.[5] This may have been associated with a change in the main source of moisture and rainfall precipitation from the Atlantic to the Indian Ocean during the onset of the major aridification

FIGURE 3.2   Climatic change as a result of the late Miocene drought.

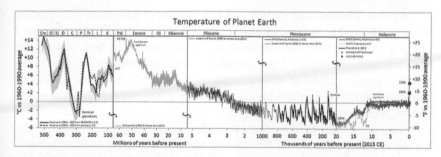

FIGURE 3.3   Temperature of the Earth over the past 60 million years. (Courtesy of Glen Fergus.)

around 8 mybp. The expansion of C4 plant life resulted from an increased equator-pole temperature gradient. The fall in global temperature during the late Miocene period from about 15 mybp (Figure 3.3) and the development of an increasing Antarctic ice sheet initiated a shift from moist woodlands with predominantly summer rainfall to a range of more arid to semi-arid environments.[6]

New sources of food had to be found, and those apes who were able to adapt more successfully would be the ones who survived and eventually would become a dominant species within that particular ecological niche. These environmental changes may have been very significant in altering the destiny of primate evolution. The traditional theory of evolution suggests that our ancestral branch of the ape family simply moved from the trees down onto the savannah, where they evolved as hunter-gatherers, standing upright because they could see further and learning 'endurance running' to be able to chase and hunt across the savannah. This concept of early hominin evolution has recently been called into question and there is increasing evidence that the savannah theory does not explain many of the unique characteristics of *Homo sapiens* when compared to other members of the ape family.

The temperature of the earth has seen some dramatic variations over the past 60 million years (Figure 3.3). Of particular interest is the fall in temperature in the late Miocene period, corresponding to the period of deforestation, commencing about 15 mybp with a dramatic fall and then slowly levelling out between 7 and 3 mybp during the Pliocene Epoch. After this, the temperature fell and fluctuated every 41,000 years or so, until about 500,000 years ago during the Pleistocene era.

We then see in the late Pleistocene a period where there were four distinct ice ages followed by a rapid rise in temperature about 15,000–20,000 years ago to a climate which has remained fairly constant since then, during the Holocene, up to the present day. Undoubtedly these fluctuations in temperature over this crucial period of several million years would have had important influences on life on earth and the evolutionary paths and migration of many species, including humans.

The changes in the temperature of the earth during and following the last ice age at the end of the Pleistocene era may well have been influential in Europe in deciding the fate of the Neanderthals and the eventual dominance of *Homo sapiens*. Neanderthals were thought to have become extinct in Europe around 30,000 years ago, but new fossils are frequently found, and the most recent Neanderthal fossils that have been unearthed in southern Spain date from about 27,000 years ago. It is unlikely that they suddenly became extinct precisely then; after all, they were interbreeding with humans and some of them may have lingered on for a while after that date in small numbers.

THE GREAT RIFT VALLEY

Apart from the dramatic changes in climate towards the end of the Miocene period, another important event occurred in this region of East Africa which would certainly have affected the 'conditions of existence' due to major tectonic activity around 6.7 mybp. The Great Rift Valley is a name given to the continuous geographic trench, approximately 6,000 kilometres in length, that runs from Lebanon's Beqaa Valley in Asia to Mozambique in south-eastern Africa. This was produced by a continental shift of the East African plate, which resulted in separation of this vast region from the main African continent. This caused inundation of a large low-lying area with formation of an extensive inland sea, the Olduvai Gorge and the Great Rift Valley. This sea eventually dried up after several million years, leaving only deep deposits of salt as evidence of its former existence (Figure 3.4).

Richard Leakey speculated that the diversity of environments resulting from the creation of the Rift Valley provided 'an ideal setting for evolutionary change'[7] which has proved to be an accurate prediction since most of the early *Australopithecine* fossils from 3–4 mybp have been found in the region of the

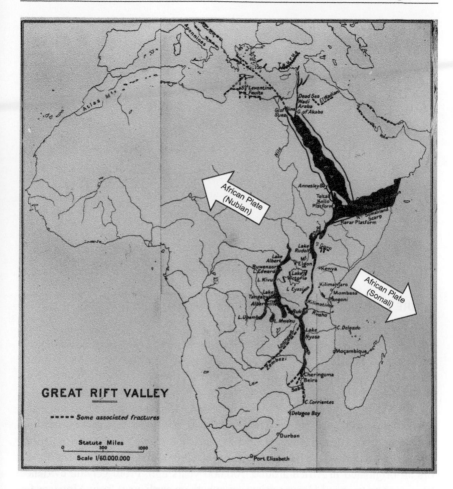

FIGURE 3.4 The Great Rift Valley. (Courtesy of the Library of Congress, Prints and Photographs Division. Reproduction number LC-DIG-matpc-00451; digital file from original photo.)

Great Rift Valley and the Olduvai Gorge (Figure 3.5), the 'Cradle of Mankind'. It is of particular interest that most of the early *Australopithecine* sites were near the branch of the Rift Valley to the east of Lake Victoria rather than the western Rift Valley. The main difference between these two branches of the Great Rift Valley is that the eastern valley consisted mainly of shallow

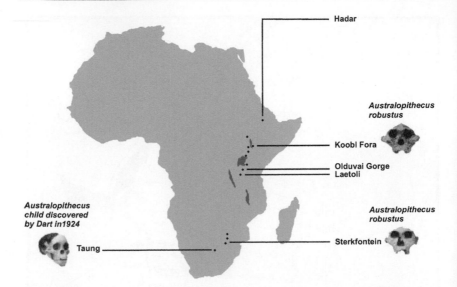

FIGURE 3.5   Early *Australopithecine* sites near the Great Rift Valley. (Courtesy of the author.)

salt water lakes, which would have provided early hominins with the essential marine habitat and diet, whereas the western valley had predominantly deep fresh water lakes.

It is clear that the early hominins evolved in Africa over several million years before the first exodus out of Africa to Eurasia and Indonesia about 2 million years ago, ending up eventually as Denisovians and *Homo floresiensis* in the Far East and Neanderthals in Europe. There is also no doubt that our human ancestors all originated in Africa around 100,000 years ago. From genetic evidence there were no traces of *Homo sapiens* anywhere on earth apart from Africa, as far as we know, and it was at about that time that they emerged in the second exodus out of Africa into Europe to join, and eventually to replace, the Neanderthals in our heritage.

As Darwin explained in *The Descent of Man*, he considered that, apart from genetic factors, 'conditions of existence' were even more powerful than 'natural selection' in determining evolutionary adaptations. It would seem that around the time of the hominid split from the ape family there were considerable tectonic and climatic changes taking place in that part of Africa that had a significant effect on the environment for animals and plants in that region.

In particular, the previously safe and stable forest habitat of the ape family was disrupted, with the loss of the usual sources of food. The increasingly arid conditions and increasing competition for the remaining woodland provided the necessary impetus for the more resourceful primates to find and establish themselves in a new aquatic habitat in the lakes and wetlands of the newly formed Great Rift Valley. In this new ecological waterside niche, the evidence suggests that over several million years they were able to evolve totally unique adaptations and characteristics of a semiaquatic mammal. Not only did they survive, but they evolved a unique mode of locomotion and lifestyle which was to enable them to become a more versatile and dominant mammal, capable of adapting and using resources in both a terrestrial and an aquatic habitat.

Our Early Ancestors

4

EARLY BIPEDAL HOMININS

The past, like the future, is indefinite and
exists only as a spectrum of possibilities.
Stephen Hawking (1942–2018)

Genetic evidence and fossil records reveal that all living humans are closely related and share a common ancestor who lived about 200,000 years ago in East Africa. The last common ancestor we shared with our closest cousin, the chimpanzee, lived about 6–8 million years before present (mybp). The first hominins appeared about this time, of whom *Sahelanthropus tchadensis*, found in Chad in 2002 and dated around 6–7 mybp, was the first to have a downward-facing foramen magnum, where the spinal cord exits the skull, indicating an erect spinal column and bipedalism. The brain capacity, however, was similar to that of the chimpanzee (Figure 4.1).[1]

Another contender for the earliest bipedal hominin was *Orrorin tugenensis*, from whom a molar tooth was found in 1974 and later a femur (thigh bone) and upper limb bones, with features suggesting bipedalism and an arboreal habitat. The shaft of the femur was thick and cylindrical, with an orientation indicating that it was adapted for bipedal walking.

Millions of years ago

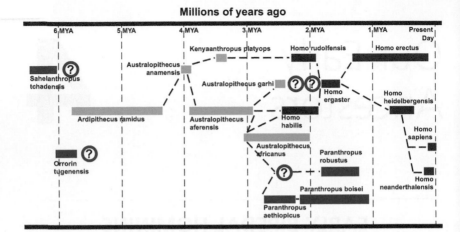

FIGURE 4.1   Humans' place in evolution. *Australopithecine* family (green); *Homo* family (blue). (Courtesy of the author.)

In Ethiopia, near the northern part of the Great Rift Valley, around 100 specimens, including a remarkably complete skeleton of another family, the *Ardepithecines*, were found in the Middle Awash River Valley sites. *Ardipithecus ramidus* remains have been dated around 4.5–4.3 million years ago, and features were suggestive of a part bipedal and arboreal habitat.[1]

AUSTRALOPITHECUS

One of the most remarkable finds in this region was an almost (40%) complete skeleton discovered by Donald Johanson in the Awash Valley in Ethiopia (Figure 4.2). 'In 1974', he writes 'I found a human-like knee joint that proved beyond doubt that our ancestors walked erect close to three and a half million years ago – long before they developed the big brains that had once been thought to be the hallmark of humanity'.[2]

Back at his base-camp that evening with the exciting new specimen, a tape recording of the Beatles' song 'Lucy in the Sky with Diamonds' was 'belting out into the night sky, played at full volume over and over again out of sheer exuberance', which is how 'Lucy' acquired her name.[2] She was undoubtedly a female because a complete pelvic bone and sacrum (the lowest bone of the spine) has been found, which also showed that the orientation of her lower limbs was bipedal, meaning that she walked erect.

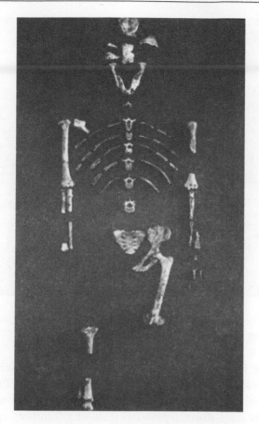

FIGURE 4.2   'Lucy' (*Australopithecus afarensis*) discovered by Donald Johanson. (Courtesy of Dr. Donald Johanson.)

A second important discovery by Mary Leakey in 1976 was the Laetoli footprints of an adult and child walking side by side, preserved in the fossilized volcanic ash, which clearly showed that these apes were bipedal (Figure 4.3). Similar discoveries of hominoid fossils in this area have proved beyond reasonable doubt that at some time between 6 or 7 million years ago and 3.5 million years ago, probably somewhere in north-east Africa, in the region of the Great Rift Valley, certain anthropoids with small, ape-sized brains, stood up and began walking erect.[3,4]

'Something must have happened', Morgan writes, 'which meant that the near universal mammalian mode of locomotion (walking on four legs) ceased to be efficient for them. They switched to a mode that was not merely different, but unique among mammals'.[5] Other proof that divergence of evolution between humans and apes from common ancestors took place about this time has come from molecular biological studies, and in particular DNA hybridization.[4]

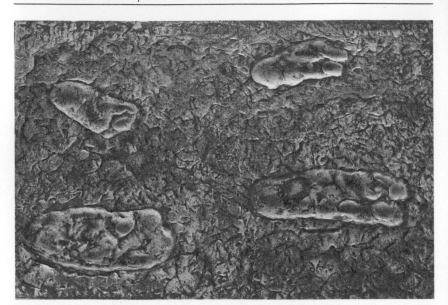

FIGURE 4.3 The Laetoli footprints. (Courtesy of Chris Howes/Wild Places Photography/Alamy.)

Much more is known from fossil evidence about hominin evolutionary changes that took place over the next few million years, although many of these lead to false routes in the evolutionary tree. We can, however, trace our ancestral lineage from *Australopithecus* (meaning 'southern ape'), who lived about 4–2 mybp through various recognizable hominin intermediaries. Apart from Lucy (*A. afarensis*), there are other branches of the *Australopithecus* family including *A. africanus* followed by *A. bosei* and *A. robustus*, *A. gigan-topithecus* ('really big ape') and *A. ardepithecus* ('ground ape'), each with different fossil characteristics (Figure 4.1).

HUMAN POPULATIONS IN THE PLEISTOCENE ERA

The Pleistocene era is defined as the period in history beginning around 2.6 million years ago and lasting until the end of the last Ice Age about 11,700 years ago. The term *Pleistocene*, from the Greek words *pleistos* (meaning 'most') and *keinos* (meaning 'recent'), was introduced in 1839 by the Scottish geologist and lawyer, Sir Charles Lyell, who was a close friend of Charles Darwin. The climate was generally much colder than it is today, and at times the average global temperature was 5°C–10°C cooler. It was also much drier because much of the water on the planet was ice and the sea levels were lower.

During the Pleistocene era, the continents had moved to their current positions, but large parts of the earth, including Antarctica and parts of South America, were covered with ice sheets. In the northern hemisphere, ice covered large parts of Europe, all of Canada, Greenland and parts of northern United States. During this period *Homo sapiens* first appeared; by the end of the late Pleistocene humans inhabited most parts of the globe.

Homo habilis

In the early Pleistocene, *Homo habilis* was associated with the next important evolutionary change. Hominin fossils found in the Olduvai Gorge by Richard Leakey from over 2 mybp included a relatively complete hand that suggested that this hominin was capable of making and using tools. Leakey's 'handyman' also had a larger brain volume of around 700 cc, still only half the capacity of *Homo sapiens* but about 50% larger than that of *A. africanus*. The earliest stone tools in the archaeological record date from about 2.6 mybp and were found in the Olduvai Gorge region of Ethiopia, where several hominin species co-existed, including both *Australopithecus* and *Homo habilis*.

Homo erectus

Within a few hundred thousand years, just before 2 mybp, the successor to *Homo habilis* emerged with a new, bigger brain. *Homo ergaster* and *Homo erectus* have long been assumed to be the species intermediate between early *Homo habilis* and *Homo sapiens*. One of the dramatic differences, however, was that the first of these *Homo erectus* fossils was first discovered in Java, Indonesia (Java Man), in 1891. It is likely that *Homo ergaster* evolved first about 2 mybp and started to extend their territory out of Africa into Asia and Indonesia by 1.9 mybp (the first hominin African exodus).

The main source of fossils, however, has been from northern Kenya, where a virtual complete skeleton of a 9-year-old *Homo ergaster* boy (some still call it *Homo erectus*), dated about 1.6 mybp, was found in 1985 in the dry bed of the Nariokotome River near Lake Turkana. He had a cranial capacity of 880 cc. The boy was more than 5 feet tall and would have exceeded

6 feet had he lived to maturity.[3,4] From studies of his teeth, some clues have suggested that he died from a dental infection leading to fatal septicaemia. His tall frame and sturdy physique seemed to be very similar to those of modern humans, prompting Alan Walker to comment: 'What sort of prehistoric basketball player had we excavated at Nariokotome?'[4] A decade earlier, an almost complete *Homo erectus* cranium was found at Koobi Fora; it had a cranial capacity of 850 cc and dated to about 1.8 mybp.

One of our ancestors' major innovations was the control of fire, which not only provided them with a means of cooking otherwise indigestible or toxic foods, but also enabled them to keep warm in the colder climates and in winter and to ward off predators. This was a major evolutionary step forward, together with tool use and a bigger brain for *Homo erectus*. It is hard to say exactly when ancestral humans began to use fire on a regular basis, but remains have been found at Lake Turkana and Swartkrans from 1.3 to 1.4 mybp which suggest that fire was used as a means of protection. More definite remains of fragments of burnt animal bones and hearths have been found in Europe and China from 300,00 to 400,000 years ago, but evidence of fireplaces on a regular basis do not appear until about 40,000 years ago. It is likely, however, that the use of fire played an important role in allowing our ancestors to extend their territory into colder and more inhospitable regions of Asia and Europe, particularly at the time of the ice ages, and also for smoking and preserving foods to help overcome seasonal shortages.[4]

Homo heidelbergensis

During the Middle Pleistocene, between 600,000 and 200,000 years ago, another distinct *Homo* species has been identified with a larger brain volume of around 1,259 cc and more advanced behaviour and tools. This species lived in Africa, Europe and Western Asia. *Homo heidelbergensis* is considered to be the ancestor of Neanderthals (European), Denisovans (found in Siberia) and modern humans.[6] The first discovery was a piece of mandible which was found near Heidelberg, Germany, in 1907, but most of the fossils from *Homo heidelbergensis* have only been found during the last 20 years in East Africa, dating from as early as 700,000 years ago.[7] Excavations have revealed that some of these skeletons were buried together, suggesting that communal burying rituals originated before the Neanderthals.

Three separate African groups have been identified, some migrating into Europe where their fossil remains have been found at sites in Spain, Germany, Italy, Greece and as far north as England.[8] A second group moved into Asia, where they evolved into Denisovans, and a third group (*Homo rhodesiensis*), who stayed in East Africa, evolved into anatomically modern humans (AMHs) between 300,000 and 200,000 years ago before migrating into Europe and Asia in the second exodus between 125,000 and 60,000 years ago.[8]

Homo floresiensis

In stark contrast to the big-brained *Homo erectus, Homo heidelbergensis* and athletic *Homo ergaster,* not far from the island where Java Man was found, fossil remains of a small hominoid were discovered on the island of Flores in Indonesia. This hominin had a small brain and was nicknamed the 'hobbit' because of its diminutive height of about 3 foot 6 inches. These fossils date from about 100,000 to 50,000 years ago, although 'hobbit' stone tools were found at the same site dating from up to 190,000 years ago. It is likely that these hominoids descended from a population of *Homo erectus* who came to Flores about 1 million years ago following the first exodus out of Africa.

They underwent a process of insular dwarfism, which results from a relative lack of food and no active predators. Despite their small stature and brain size, these early humans hunted small elephants and large rodents and probably used fire. Their demise seems to have coincided soon after the arrival of *Homo sapiens* in Indonesia, which suggests that they too may have undergone the same fate as the Neanderthals in Europe 20,000 years later, who became extinct 30,000 mybp following the second exodus of *Homo sapiens* from Africa into Europe around 60,000 years ago.

Homo denisova

In south-western Siberia, in the Altai mountains near the border of China and Mongolia, an important discovery was made of a cave which is named after Denis, an eighteenth-century Russian hermit. This cave was explored originally in the 1970s. In 2008, Michael Shunkov and other Russian archaeologists did further excavations and found a finger bone of a juvenile hominin. Other artefacts found in the cave (including a bracelet) were dated to around 40,000 years ago, but recent findings of objects from that site reveal that these relics go back as far as 125,000 years ago, and include relics of the branch of *Homo heidelbergensis,* who ventured out of Africa into Western Asia, and also *Homo neanderthalensis.*

The cool climate of the Denisova cave preserved the ancient DNA, and analysis has amazingly revealed that these Denisovans, as well as the European Neanderthals and modern humans, all shared a common ancestor around 1 million years ago (Figure 4.4). The work of Rogers and colleagues from Stanford University has revealed that during the late Pleistocene, the

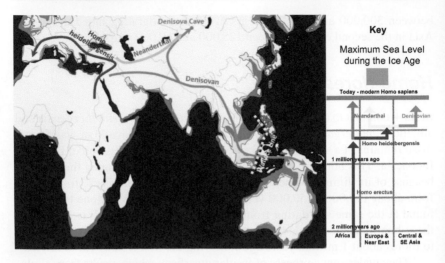

FIGURE 4.4   The evolution and spread of Denisovans and other hominins. (Courtesy of the author.)

Neanderthals and Denisovans separated only a few hundred generations after their ancestors had split from AMHs.[9]

Many different theories and models attempt to explain how our early bipedal hominin ancestors evolved in East Africa and later spread to other parts of the globe. Archaeological fossils and specimens provide the essential evidence to help identify changes in appearance, use of tools and social behaviour, but genetic data is increasingly vital in determining the relationship between various hominin species.

The
Neanderthals
and Their
Demise

5

*The human failing I would most like to correct is
aggression. It may have had a survival advantage
in caveman days to get more food, territory or
a partner with whom to reproduce, but now it
threatens to destroy us all.*
Stephen Hawking (1942–2018)

Perhaps one of the most well-known of our ancestors is the Neanderthal, or more correctly *Homo neanderthalensis*, whose fossilized skeletal remains from about 40,000 years ago were originally found by quarry miners in the Neander valley in Germany in 1856, just a few years before the publication of *Origin of Species*. Large numbers of Neanderthal skeletons have since been found throughout Europe, the Middle East and parts of Asia where they lived between 400,000 to around 30,000 years ago, towards the end of the Pleistocene Ice Age. Many of these skeletons and remains are in excellent condition and well preserved, some laid together in groups, suggesting that these early hominins were intentionally burying their dead. Evidence suggests that the Neanderthal population grew large and fragmented into isolated groups.[1]

Neanderthals certainly received bad press after their discovery for a long time. During the Victorian and Edwardian periods, they were often depicted as vicious hairy savages (Figure 5.1). There was even a suggestion that they should be named *Homo stupidus*. *Homo sapiens* was considered to be the pinnacle of human evolution and any other potential primitive contender for a place in our family history was regarded with great scepticism.[2] In a century concerned with the developments of science and technology, within which the boundaries of science were pushed far more than ever before, it is

FIGURE 5.1   Victorian image of the Neanderthal. (Courtesy of Chronicle/Alamy.)

initially surprising that so little attention to the primitive mind could be seen in British scientific circles.

Anatomically, Neanderthals have several distinctive features which are evidence of their geographical distribution and the cold climate in which they lived. Their skeleton is short and robust with heavy musculature, which would help minimize heat loss through the skin. In contrast to humans who live in hot climates, who are usually tall and thin, those living in cooler climates, such as the Eskimos, are of shorter stature. This is demonstrated by comparing the length of the tibia (shin bone) with the femur (thigh bone). The ratio between these two measurements is called the Crural Index, and in Neanderthals it has been found to be low, whereas in early modern humans the ratio is quite high.[3] Some of the Neanderthals remains found in the warmer climates of eastern Europe and western Asia, however, had lighter skeletons.

One of the other striking features of Neanderthals was the size of their brain, which was as large, measuring about 1200–1750 cc in adults, or even larger than those of modern humans (Figure 5.2). Their skulls were broader but had a flatter brow area and are also characterised by a prominent mid-facial structure with an elongated nasal skeleton. This may have been beneficial in increasing the size of the nasal cavity to help warm the inspired air in their cold environment. There is

FIGURE 5.2   Skulls of Neanderthal Man (La Ferraissie) and anatomically modern man (Cro-Magnon). (Courtesy of Science Photo Library/Science Source.)

also strong evidence from the comparative size of their cerebral hemispheres that they were mainly right-handed, similar to modern humans.

Chris Stringer, a renowned paleontologist from the Natural History Museum in London, is credited with detailed analysis of contemporary Neanderthals and Cro-Magnons and for introducing the 'Out of Africa' theory.[4] Cro-Magnons were the first anatomically modern human (AMH) in Europe since the shape of their skulls was much more similar to that of modern humans compared to that of Neanderthals (Figure 5.2). One of the important features that Chris Stringer identified was that the Crural Index in Cro-Magnons was similar to that of modern humans and much higher than that of Neanderthals. He found no evidence that Neanderthals had evolved into modern humans and firmly believed that AMHs first evolved in East Africa around 200,000 years ago and eventually migrated out of Africa (the second exodus) sometime during the last 100,000–125,000 years. It is thought that they arrived in Europe about 60,000 years ago, by which time the Neanderthals were already well established there, although in small numbers. Thirty thousand years later, however, *Homo neanderthalensis* was extinct.

There is much controversy about the fate of the Neanderthals in Europe and why there are no significant remains of *Homo neanderthalensis* more recent than 30,000 years ago in France, although they seemed to have survived in southern Spain and Gibraltar until about 27,000 years ago.

In 1868, some of the first bones of the AMHs were found in the Cro-Magnon cave near the French town of Les Eyzies, dating from around 40,000 years ago. Cro-Magnons were the first of the *Homo sapiens* people to be found in Europe and the fact that they lived at the same time and in the same region of France as the last of the Neanderthals has intrigued scientists for over a century. Even more important is trying to discern the relationship between these two hominin populations.

Although, in evolutionary terms, the last common ancestor of *Homo sapiens* and *Homo neanderthalensis* was more than half a million years ago, genetic evidence has now shown convincingly that when the new *Homo sapiens* arrived in Europe from Africa several hundred thousand years later, the two populations and tribes were sufficiently similar to interbreed and leave their combined DNA sequences for genome analysis.

About 2–3% of the DNA of typical modern Europeans is derived from Neanderthals, but in some eastern Asians there is a higher concentration of Neanderthal DNA, and most Africans have very little. DNA extracted from Neanderthal bones found in Croatia showed that the two populations had interbred around 60,000 years ago, and in 2014 similar findings of mixed AMH and Neanderthal DNA have been made in Siberia from a female who died 50,000 years ago. More detailed analysis of her genome in 2016 revealed that the genetic introgression (fusion of DNA from two separate groups) had occurred in one of her ancestors around 50,000 years before she was born, which suggests that the second exodus of AMHs from Africa to Europe and Asia may have occurred around 100,000 years ago, earlier than had been believed at the time.[5]

So what happened to the Neanderthals who had survived in Europe and Asia through much of the last ice age and yet were replaced within a relatively short evolutionary time by a new breed of hominins who had evolved in a hot tropical climate? It is clear that the two evolutionary cousins lived together in supposed harmony for them to interbreed, and genetic evidence implies that the first mixed offspring were the result of an encounter between a male Neanderthal and a female *Homo sapiens*.[6]

Several theories have been offered to explain the demise of the Neanderthal family. They had big brains and strong physiques, but it is possible that the AMH newcomers were much more versatile and creative, as exemplified by the findings of Cro-Magnon art in a cavern in the Ardeche Gorge in south-eastern France dated around 31,000 to 33,000 years ago (Figure 5.3). It seems that around 40,000 years ago there was a cultural revolution in *Homo sapiens* associated with the advent of Aurignacian creativity, during which many more

FIGURE 5.3   Cro-Magnon art from Ardeche Gorge, France. (Courtesy of thipjang/ Shutterstock.)

sophisticated tools including bone spearheads, fishhooks and harpoons were found.[7] The Neanderthals were associated with Mousterian technology which showed little use of bone, antler and ivory and, whereas the raw materials for making their tools and implements were mainly from local sources, those associated with *Homo sapiens* were derived from much further afield, indicating a more diverse culture and trading.

However, the recent publication in *Science* about cave art in La Pasiega and two other sites in Spain suggest that these ochre pigment paintings were created by Neanderthals at least 64,000 years ago, which is at least 20,000 years before the supposed arrival of *Homo sapiens* in Europe.[8] This artwork is very similar to paintings found in Sulawesi, an Indonesian island, and elsewhere in Europe from around 40,000 years ago.

Evidence suggests that differences in the ability to communicate, including speech and a higher symbolic intellect and consciousness, may have been influential factors in the supremacy of *Homo sapiens*, or it may just have been a question of greater numbers. It is also possible that *Homo sapiens* introduced new diseases or infections from Africa to this mixed European community and that the Neanderthals did not possess the necessary immunological defences to combat them. In recent history we have seen how the plague, which ravaged Europe in the seventeenth century, killed one third of the population in

England and elsewhere, and it wasn't very long ago that the acquired immune deficiency syndrome (AIDS) epidemic, which started in Africa, led to millions of deaths worldwide.

However, the answer may simply be that *Homo sapiens* was able to survive better in the same small ecological niche that they shared with the Neanderthals in Europe at that time. Neanderthals had evolved and survived in Europe during the difficult time of the last four ice ages as sturdy muscular humans who were able to conserve heat and were adapted to those harsh conditions. Although there is some evidence that the second exodus of AMHs from Africa was as early as 100,000 years ago,[4] most scientists believe that AMHs arrived in Europe towards the end of the last Ice Age, around 60–70,000 years ago. As the ice cap began to recede, resulting in a warmer climate, their more agile and taller AMH ecological rivals, whose ancestral pedigree was used to a subtropical climate in East Africa, may have been better adapted to survival in this less hostile environment. These Cro-Magnon humans were also possibly better suited to running in a more open and drier terrain, and more successful in hunting, swimming and diving, particularly with their more sophisticated arsenal of weapons.

Some other large mammals, such as the woolly mammoth, giant bears and the sabre-toothed tiger, which had adapted successfully to the ice-age conditions, became extinct around 13,000 years ago. Scientists have long debated the cause of their rapid demise, whether it was from overhunting by humans or an inability to survive the harsh conditions. But recent evidence suggests that their sudden extinction might have been the result of a major cosmic disaster, such as the large comet which exploded over southern Canada, causing massive global cooling over 1,300 years. Fossils from around 35 different mammalian species were no longer found after this date, and it also seems to be associated with the end of the human North American Clovis culture.

Many of the mammals that we recognize today, including primates, cattle, deer, members of the dog and cat families, as well as kangaroos, wallabies and bears, adapted and thrived during the Pleistocene era. As the earth grew warmer and the ice fields retreated, the water levels rose; the previously dry, exposed continental shelf areas of Europe and North America were inundated, and estuaries and rivers filled and widened. The British Isles became separate from mainland Europe, and the shapes of the continents became more similar to what they are today. This increasingly aquatic environment may also have been much more suited to a slender swimming and diving hominin who had evolved over several hundred thousand years with all of the unique additional characteristics of a semiaquatic mammal that we see in modern *Homo sapiens*.

The current Holocene epoch sees the emergence of modern humans and the age of agriculture and domestication of animals about 10,000 years ago. Whatever the reason for the demise of the Neanderthals, they are part of our genetic ancestral heritage, and we have the indelible imprint of their DNA in our genome, analysis of which is rapidly advancing our knowledge and understanding of the true and complete story of hominin evolution.

The Waterside Ape

6

Why Are We So Different?

*We have to admit to being baffled about the
origin of upright walking. Probably our thinking
is being constrained by preconceived notions.*
Sherwood Washburn and Roger Lewin

There is general agreement that around 7 to 9 million years ago our ancestral ape family members lived with other quadruped apes in the forests of East Africa, having evolved in this habitat over many millions of years, well adapted to their environment. Over the next 2 to 3 million years, however, because of external environmental factors which adversely affected their status quo, some family members of this last common ancestor were to undergo further evolutionary changes resulting in a divergence or split, which led to the emergence of apes with some different characteristics. Each branch of the family was to go down different evolutionary pathways, adapting to a slightly different ecological niche or habitat in which they became familiar and found ways of surviving.

As Jacob Bronowski summed up in his *The Ascent of Man* television series: 'Human evolution began when the African climate changed to drought: the lakes shrank, the forest thinned out to savannah'.[1] The eventual outcome saw the emergence of the gorilla, chimpanzee, bonobos and humans, who survive today (Figure 6.1).

The story of human evolution does not fully explain what happened at that time or find a reason why our other primate cousins remained largely unchanged over the past 10 million years to the present day, and yet we humans have so many different features and characteristics that are totally unique among other primates and indeed among any other terrestrial mammals.

FIGURE 6.1   The three African ape cousins: Why are we so different? (Left: Courtesy of Michael Poliza/National Geographic Creative. Right: Courtesy of Air Images/Shutterstock.)

Some other primate species such as the vervet and patas monkeys and the baboons came down from the trees to live on the savannah millions of years ago, and yet they show no sign of acquiring hominin features such as bipedalism or hairlessness and have remained much the same with scarcely any physical modification until the present day.

Some of the unique terrestrial mammalian charcteristics of humans, however, as observed by Hardy,[2] are seen commonly, in various forms and combinations, in aquatic and semiaquatic mammals. Each of these human characteristics on their own are not individually exclusive proof of an aquatic or semiaquatic environmental influence in our evolution, but put together, there is now overwhelming scientific evidence that at some stage during the past few million years, early hominin humans had a substantial influence of a littoral (waterside) habitat which has resulted in the emergence of a dominant, intellectual, large-brained species, *Homo sapiens*. As Professor Philip Tobias, one of the most pre-eminent paleoanthropologists concluded, these evolutionary changes could not have taken place on the savannah, as traditionally thought, but must have involved a significant waterside influence in the daily life of early man.

When studying the comparative anatomy and physiology of primates and other mammals, one notices that there are a number of unique human characteristics which are not seen in any of the other terrestrial species, and these anomalies are difficult to explain in the context of evolution of hominin as a terrestrial mammal on the savannah. These do make complete sense, however, when compared with evolutionary adaptations in other semiaquatic species over several million years. Apart from general characteristics which are mentioned here, there are other specific features relating to the upper aero-digestive tract affecting our breathing, hearing and voice, as well as brain function, which are discussed in subsequent chapters, which add further evidence to the waterside ape theory of evolution.

BIPEDALISM

Since the time of Darwin, one of the most important questions in human evolution was, What came first, the bigger brain or bipedalism? Did early humans initially develop a bigger brain with more intelligence and then decided to stand upright because it would enable them to run and hunt, freeing their hands to be able to develop better manual dexterity for holding weapons and carrying food? Or did they become bipedal initially, unlike their cousin apes, and then realise that their arms and hands were free to use for these purposes? The answer came with the finding of Lucy and the Laetoli footprints in the 1970s, which proved that bipedalism came first, because the bipedal *Australopithecines* still had chimpanzee-sized brains.

But why should one branch of the ape family decide to stand upright? After all, of over 200 species of primates, only one is bipedal; of over 4,000 species of mammals, only one is bipedal. That exception is the human, the hairless bipedal ape. Something must have happened to initiate this major unique evolutionary change in one branch of our ancestral family. Quadruped gait is the almost universal means of locomotion for ground-dwelling mammals and has evolved over 50 million years; it provides stability, is efficient in terms of energy and is easily learned in the new-born to afford independence within a few days of birth. The strong horizontal spine provides support for the vital internal organs in the chest and abdomen.

No animal could afford to sacrifice all of these assets without an overriding powerful selective pressure.[3] The only other mammals who can successfully adopt a bipedal gait, albeit hopping, are the kangaroo and the wallaby who have a strong stabilising tail for additional support.

NAKEDNESS

The loss of hair in humans has been difficult to explain in evolutionary terms, since a covering of fur or feathers has been fairly standard in warm-blooded creatures on land ever since they emerged over 100 million years ago, to help regulate their body temperature. Even Darwin found it difficult to explain: 'The loss of hair is an inconvenience and probably an injury to man, for he is thus exposed to the scorching of the sun and to sudden chills, especially due to wet weather. No one supposes that the nakedness of the skin is any direct advantage to man; his body therefore cannot have been divested of hair through natural selection'.[4] The topic of nakedness is covered in the next chapter.

SUBCUTANEOUS FAT

Land dwellers have relatively thin skins. The subcutaneous fat between the subcutaneous muscles (platysma) and the muscles of the body is rarely a continuous thick layer.[5]

V. E. Sokolov (1982)

There are two types of mammalian fat – brown fat and white fat. The former has a richer blood supply, which gives it a darker colour, and is found predominantly in young mammals, where its function is to respond to a drop in body temperature and provide rapid heating. In a human neonate, fat constitutes 16% of body weight compared with 3% in the baboon. White fat is mainly seen in mature mammals and is used as a store of energy in terrestrial mammals rather than for insulation. In hibernating mammals such as bears, hedgehogs, dormice and marmots, which live in cold climates where food is scarce in winter, their fat layer is seasonal.

Humans are unique among primates and among most land mammals in their propensity to accumulate comparatively large amounts of adipose tissue, especially in subcutaneous sites.[3] This would seem to be disadvantageous in a savannah environment where the additional weight would slow down the hunter-gatherer ape.[6] In a comparison of 23 mammalian species ranging from bats to whales, it was found that humans have at least 10 times as many adipocytes (cells that produce fat) as would be expected in proportion to their body weight.[7] It is difficult to establish whether the *Australopithecines* and other early hominins had evolved this increase in adipocytes or whether it evolved at a later date because, like skin, these are soft tissue changes and are not preserved as fossils.

It has been suggested that accumulation of fat in humans has come as a result of the advent of an agricultural and more settled lifestyle, 8000–10,000 years ago, but this is unlikely since these ingrained human features would need a much longer period of evolution. Also, those extant humans practising a non-agricultural lifestyle have babies with an equally plump appearance from birth. The Venus figurines found in several sites in Europe, some of which date to more than 30,000 years ago, would also confirm that this generous distribution of fat in humans, at least in females, has been established early in hominin history, long before the advent of an agricultural way of life (Figure 6.2).

FIGURE 6.2   Venus figurines dating from 30,000 years ago. (Courtesy of Natural History Museum, London/Bridgeman Images.)

THERMOREGULATION

The more you cry, the less you pee.

Folk saying

The maintenance of a fairly constant body temperature in mammals is essential for optimum and efficient metabolism of bodily functions. In human populations in different parts of the world, where ambient temperatures vary enormously, the shape of our bodies does significantly adapt to the climate in which we live. This variation has long been recognised and certain rules have been established to explain the general principles relating to body shape and size. Bergmann's rule, published in 1847, states that for any geographically widespread species, those populations in warmer climates will have a smaller body (trunk) than those in colder regions. Allen's rule, from 1877, relates to the limbs and states that those living in warmer climates have longer arms and legs than those from colder regions.[8] These rules are generally recognised in comparing the short, stocky Eskimos with the tall Nilotic people of Africa and are associated with the balance of heat production and the ability to dissipate it.

More recently, Ruff has studied various archaeological and recent populations and has found that these predictions are sustained very well.[9] Humans rely heavily on sweating to cool their bodies, and in open environments in Africa Nilotics can efficiently lose excess heat by this method. Mbuti Pygmies, on the other hand, live in moist, humid forests where evaporation of excess surface fluid is inefficient and they cannot rely on this to reduce body temperature. As a result, the most efficient strategy is to reduce the size of the body with a smaller stature. This climatic adaptation can also be seen in the Neanderthals who lived in Europe between 300,000 and 27,000 years ago during the late Pleistocene Ice Age, where their small, stocky build, like the Eskimos, reduced the amount of heat loss.

Sweating seems to have appeared at a relatively late stage in mammalian evolution. Smaller mammals, including chimpanzees and gorillas, pant when they get too hot and moisture is lost from the lungs through increased respiratory effort, which seems to be quite efficient as a method of cooling. Humans appear to be the only land mammal which does not pant as a means of cooling down; we do pant during exercise to breathe in more oxygen but we do not pant when lying in the sun.[6]

Larger mammals that spend a lot of time out in the open, such as cattle, horses, sheep and other grazing animals, have developed the ability to sweat in addition to panting as a means of cooling down. Sweat glands on the skin surface occur only in mammals and are of two types. *Apocrine* glands are associated with hair follicles and were probably designed originally to secrete an oily substance for lubrication of the hairs and also for giving off scent recognition. In most land mammals, including primates, with the one exception of humans, these glands are found all over the body and when the animals' body temperature rises, they secrete a diluted watery emulsion, including salts, which evaporates on the skin to produce a cooling effect. Excretion of water and important salts such as sodium chloride and potassium are efficiently controlled and salts can be reabsorbed.

In humans, these apocrine glands are present prenatally to about the fifth month, but after birth they are only found under the arms, in the groin area, around the nipple and in the navel where they do retain their scent-producing function. They only become active at puberty and their secretions are largely stimulated by emotional and sexual excitement. In the axilla these apocrine glands in higher primates, including humans, are known as the axillary organs, and their secretions can become odorous because of decomposition by bacteria when they are on the skin. Some apocrine glands are specialized to produce wax in the ear canal, and mammary glands are thought to be highly modified apocrine glands.

But what about the near universal function in land mammals and primates of apocrine glands acting as an efficient means of thermoregulation? In humans, apocrine glands are too few in number to be adapted for temperature control, and it does seem strange that in land mammals, uniquely seen in *Homo sapiens*, these apocrine glands have become obsolete for no apparent physiological reason and the temperature regulation function is instead taken over by millions of *eccrine* glands. These glands open directly onto the surface of the skin and are functioning from birth, secreting a clear watery fluid which contains salts.

The eccrine sweat glands are controlled by the *sympathetic nervous system*, and when the body temperature rises, the eccrine glands are stimulated to secrete a watery fluid onto the skin surface, where heat is removed by evaporation, producing a cooling effect. In humans, the rate of sweat secretion is higher than that of any other known mammal, and these eccrine glands are therefore the major means of thermoregulation. They are active over most parts of the body, with the highest density on the palms of our hands and the soles of our feet and then on the head.

Eccrine sweating has the disadvantage, however, that its onset is delayed between 5 and 30 minutes from the initiation of the response to a rise in temperature, which might initiate sunstroke. Another disadvantage is that the loss of water and salts from the skin continues, even after dehydration and/or salt depletion are reaching dangerous levels.[3] Sweat may therefore be secreted more quickly and profusely than necessary for the thermo-regulatory benefit and the additional loss is wasted. Research has shown that soldiers engaged in strenuous activity in a hot climate may lose up to 10 to 15 litres of water per day through their skin; unless it is replaced quickly, it may prove fatal,[6] as it did tragically with soldiers exercising in the Brecon Beacons 4 years ago on the hottest day of the year. This is why in humid conditions sweating affords little or no relief.

In primates and in all non-primate species such as dogs, cats, cattle and sheep, the eccrine glands are only found on the pads of the paws where they help to improve grip. In many of these animals, temperature control is by panting. The effect of improving grip can be demonstrated in humans when we moisten our fingers to turn over the page in a book. Smaller mammals, such as rodents, cannot endure dehydration and hence possess no eccrine glands at all. In primate evolution, however, eccrine glands appear and in the African ape they are as plentiful as apocrine glands, but in humans, the ratio between eccrine and apocrine glands is about 99–1%.

In human evolutionary terms 'sweating is an enigma that amounts to a major biological blunder', writes William Montagna, 'it depletes the body not only of water but also of sodium and other essential electrolytes that are carried off with the water'.[10] Vladimir Sokolov, who also spent his life studying mammalian skin, could not offer an explanation for eccrine sweating in humans, which 'is a human characteristic as unique as speech or bipedalism'.[5] However, like other scientists and anthropologists of his era, he assumed the traditionally accepted savannah theory that hominins evolved as a terrestrial mammal.

There does not appear to be any physiological reason why the efficient mammalian apocrine thermo-regulatory control in land-based primates should be uniquely altered in humans to an inefficient eccrine gland system with unnecessary wastage of water and salts, unless the 'conditions of existence' promoted another vital and different pathway of hominin evolution from our African ape cousins. Eccrine salt-excreting glands are only appropriate in a cool, wet, briny environment and would have been potentially fatal in terrestrial mammals.

On the savannah where water is not readily available, it is extremely unlikely that a totally different eccrine sweat-cooling system could have evolved in a branch of the ape family. Unlike the camel, which can drink 30% of its body weight without stopping, which sustains it for a considerable period of time, we humans are dependent on frequent rehydration, especially for hunters in a subtropical climate where our heat loss is considerable. The dependence on water and salt in early hominins would limit the range for foraging and hunting.[11] Apart from water depletion, the other major concern is loss of precious salts in sweat. It has been estimated that, at maximum sweating capacity, the body can lose its entire sodium reserve in just three hours, with potentially fatal consequences.[6]

In light of this evidence, early hominin evolution on the savannah is a most unlikely scenario and the alternative explanation that early humans evolved in a waterside habitat with a plentiful seafood diet is much more compelling and physiologically much more logical.

BIG BRAINS

One of the most compelling arguments supporting the waterside aquatic theory is the question of our bigger brains and higher intelligence, which is explored more fully in later chapters. Compared with our primate cousins, gorillas and chimpanzees, humans have a much larger brain and far higher intelligence compared to our body size. Aquatic and semiaquatic mammals, however, like the dolphin and whale, are much more intelligent than terrestrial mammals, and we frequently hear of new evidence of their extraordinary intellectual behaviour.

This difference in brain size and intelligence in mammals between those living on the land and those with a semiaquatic or aquatic existence would seem to be due to a difference in diet. There is conclusive evidence that mammals having the benefit of a marine or aquatic diet have an abundant supply of two essential lipoproteins – docosahexaenoic acid (DHA) and arachidonic acid – that are essential for brain enlargement and nerve regeneration. These are not readily available in the terrestrial food chain, and it would therefore not have been possible for early hominins living on the savannah to have acquired a large brain or to have evolved a significantly higher intelligence than their cousin apes.

Over the last 150 years since the time of Darwin, there have been so many proposed theories to try and explain the many substantial and unique differences in physiology, anatomy, behaviour, intelligence and lifestyle between hominins and our closest primate cousins. None have been at all convincing, nor has there been any evidence of a similar parallel path of evolution in any other terrestrial mammal. The simple reason is that all of these theories have been based on the unsubstantiated assumption that early hominins evolved from our primate lineage by making the transition from a quadruped arboreal ape to a bipedal hominin on the grassland and savannah below. The most obvious differences between hominins and our ape cousins are highlighted in this chapter, but many other evolutionary adaptations, such as acquisition of speech, marine-type kidneys, ear exostoses, enlarged sinuses and neonatal differences, add further convincing evidence of the waterside theory.

The Naked Ape

7

> *There are 193 living species of monkeys and*
> *apes. One hundred and ninety-two of them are*
> *covered with hair. The exception is a naked*
> *ape, self-named* Homo sapiens.
> Desmond Morris (b. 1928)

One of the most conspicuous differences between humans and other primates, the monkeys, apes, and in fact most terrestrial mammals, apart from being bipedal, is our nakedness and lack of body hair. The reason why we humans lost our hair and at what stage in our evolution from the apes this happened have been the subject of much speculation and many different theories. These have been based mainly on the traditional theory of early hominins moving from the trees onto the savannah below following the split from our arboreal quadruped cousins around 6–7 million years ago. Recent evidence has suggested, however, that the major differences in our skin, hair and subcutaneous tissue compared with our ape cousins may have come about not because we adapted differently to other apes on the subtropical savannah to become bipedal hunter gatherers but because our ancestral ape family embarked on a totally different course of evolution involving a semiaquatic habitat rather than the traditionally held idea of a terrestrial savannah environment.

STRUCTURE AND FUNCTION OF THE SKIN

The skin, which forms a protective covering over the surface of an animal, is the largest organ in the body and serves many purposes. In humans and other mammals, functions of the skin include the following:

1. *A protective barrier.* It provides a physical waterproof barrier that protects the body from mechanical, chemical and microorganism injury.

2. *Temperature regulation.* The skin helps to maintain a constant body temperature through variations in the superficial blood circulation, sweat production and the presence of hairs and, in humans, an insulating layer of subcutaneous fat.
3. *Immunological protection.* The skin provides an important barrier which senses and responds to external stimuli, without which the skin is prone to infection, inflammation, allergy and skin cancer.
4. *Sensation.* Sensory nerve endings for touch, temperature and pressure in the skin provide a constant way of monitoring the external environment.
5. *Radiation protection.* The melanin pigment in the skin provides protection against potentially damaging ultraviolet light radiation.
6. *Injury repair.* Following injury to the skin, a process of repair is initiated which attempts to heal the wound and restore skin integrity.
7. *Appearance.* The skin is draped over the underlying skeleton, muscles and soft tissues of the body and provides the recognizable and distinguishing features of the animal, which are constantly changing with the aging process and other internal and external influences.

The basic structure of the skin in mammals is fairly uniform, but clearly it has evolved depending on the immediate environment in which the animal lives, to which it has to adapt. The habitat may be terrestrial, as it is for our immediate primate cousins and other land-based mammals, or it may be purely aquatic, as it is for dolphins, whales and other cetacea. For humans and other semiaquatic mammals, the skin has had to evolve to accommodate both a terrestrial habitat and an aquatic environment, which means that many of the anatomical and physiological features are a compromise of some characteristics that are seen in either aquatic mammals or land-based creatures. Sir Alister Hardy was a marine biologist who first noted the similarities between human and marine mammalian skin, and pointed out the obvious differences between us and our primate cousins, and it was he who really started the interest in the aquatic ape theory in 1960.

The skin is composed of two main layers: the epidermis, made of closely packed epithelial cells which provide protection and waterproofing, and the dermis, made of dense, irregular connective tissue that houses blood vessels, hair follicles, sweat glands, and other organs (Figure 7.1). These structures in the dermis include the tiny arrector (or erector) pili muscles which are attached to the stem of each hair or feather to vary its direction and height. In humans, when the muscle contracts, the hair becomes erect, causing goose bumps, which is part of the fight or flight reaction mediated by the sympathetic autonomic nervous system. This means that it is subconsciously controlled and it is important as part of the thermoregulation system to trap air next to the skin so that mammals can increase the thickness of their coat and layer of insulation when cold, or bristle

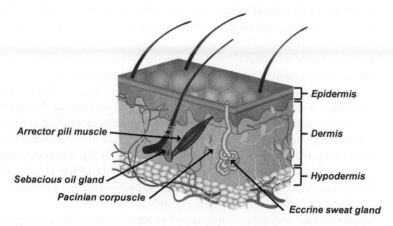

FIGURE 7.1 Typical mammalian skin structure. (Courtesy of the author.)

their hairs when in danger to appear larger and more aggressive. This is particularly critical in some mammals, especially in those with thick coats living in cold climates. A fur or hairy coat is therefore the first line of defence against heat and cold and also against potential damage from ultraviolet rays from the sun.

In primates and other terrestrial mammals, a dense hair coat seems to be fairly uniformly distributed and of similar consistency over most of the body, except on the ventral (under) surface of the abdomen where the hair is thinner. In humans, the skin also appears to be fairly uniform over the whole body, but the hair is short and fine, apart from some areas where the hair is more conspicuous, for instance on the scalp, under the arms and in the groin. In females the hair is much less evident. This nakedness has evolved because of our different semiaquatic lifestyle and, as a result of losing our hair, we had to evolve a different means of thermoregulation in the skin. The arrector pili muscles have therefore become more vestigial (obsolete) in humans (we can now buy a nice thick fur coat to keep us warm!).

APOCRINE GLANDS

Attached to the shaft of each hair is also a small apocrine sebaceous gland which secretes an oily substance through a narrow duct to lubricate the hair. Most other mammals have numerous apocrine glands in the hairy skin. These glands are especially well developed in semiaquatic mammals such as the beaver and otter, who spend much of the time in water where the oily coat helps to reduce surface tension and aid streamlining when swimming. They need their thick coat because they are exposed to a wide range of temperatures in their particular habitat.

Apocrine glands in humans have nothing to do with sweating. They appear late in fetal development (5 to 5½ months) nearly everywhere on the body. Most of these rudiments disappear within a few weeks except in the external ear canals where they specialize to produce wax; in the axillae; on the nipples of the breasts; around the navel; and on the anogenital surfaces. Single glands may be found anywhere. From this, one might speculate that the ancestors of humans had apocrine glands widely distributed over the body, like our primate cousins, and the embryonic remnants may be reminders of the history of a once widespread system of these glands, which then disappeared with the loss of hair, associated with our semiaquatic lifestyle.

The odour of axillary secretion becomes more intense as it is decomposed by bacteria. Although axillary odours frequently seem unpleasant, they are not invariably so. The odour of individual human beings comes mostly from apocrine secretion, with some contribution from sebum. Since the body odours of all other animals have a social or sexual significance, it can be assumed that this is the archetypal purpose of apocrine secretion, even in humans. The view that the axillary organs are scent glands is supported by the finding that androsterones – the compounds that are responsible for the odour of the boar to which the sow responds – also occur in human axillary secretions.

In humans these sebaceous glands can be the source of much embarrassment and discomfort because the glands may get infected, causing acne, or the ducts can get blocked, which results in swelling of the glands to produce sebaceous cysts or abscesses when they become infected. These are particularly common on the scalp, on the face and neck and under the arm, but treatment is fairly straightforward with excision of the cyst. Acne is essentially a hormone- and age-related problem for which there are effective medications. It is usually self-limiting, although the resulting scarring of the skin can be distressing.

ECCRINE GLANDS

Eccrine glands are usually absent from the hairy skin and are limited to friction surfaces (Figure 7.1). In nonhuman primates, there is a tendency for the number of eccrine sweat glands over the body to increase in progressively more advanced animals at the same time that the number of apocrine glands becomes reduced. Prosimians (primitive primates, such as lemurs, lorises and tarsiers) have only apocrine glands in the hairy skin; eccrine glands began to appear in some of the higher forms. The great apes either have equal numbers or have more eccrine than apocrine glands. Humans have the most eccrine glands, with apocrine glands restricted to specific areas such as the scalp, groin and axillae.

Humans have 2,000,000 to 5,000,000 eccrine sweat glands, with an average distribution of 150 to 340 per square centimetre. They are most numerous on the palms and soles and then, in decreasing order, on the head, trunk, and extremities. Some individuals have more glands than others, but there is no difference in number between men and women.

The specific function of sweat glands is to secrete water on to the surface so that it can cool the skin when it evaporates. The purpose of the glands on the palms and soles, however, is to keep these surfaces damp, to prevent flaking or hardening of the horny layer, and thus to maintain tactile sensibility. We moisten our fingers to turn over the page of a book; a dry hand does not grip well and is minimally sensitive.

The eccrine glands then can be divided into those that respond to thermal stimulation, the function of which is thermoregulation, and those that respond to psychological stimuli and keep friction surfaces moist. This makes a clear-cut distinction between the glands on the hairy surfaces and those on the palms and soles. In addition to thermal and psychological sweating, some individuals sweat on the face and forehead in response to certain chemical substances.

The eccrine glands on the palms and soles develop at about 3½ months of gestation, whereas those in the hairy skin are the last skin organs to mature, appearing at five to 5½ months, when all the other structures are already formed. This separation of events over time may represent a fundamental difference in the evolutionary history of the two types of glands. Those on palms and soles, which appear first and are present in all but the hooved mammals, may be more ancient; those in the hairy skin, which respond to thermal stimuli, may be more recent organs. The skin of monkeys and apes remains dry even in a hot environment. Profuse thermal sweating in humans, then, seems to be a new function, evolved because of our semiaquatic lifestyle, and is essential for keeping the human body from becoming overheated.

WRINKLY FINGERS

One of the most curious things about having a nice warm bath is that after a while, the tips of our fingers become rather wrinkled, resembling the surface of a dried prune. This can happen between 5 and 30 minutes after immersion in water. Research has shown that, like the contraction of the arrector pili muscles of the skin, this is under the control of the subconscious sympathetic autonomic nervous system. There have been many theories about why this happens, and in the past people have thought it was due to the skin soaking up water or due to nerve damage. However, it has been known for some time that if a nerve to a finger is cut, that finger will no longer go wrinkly, which confirms the neural control.

In 2011, a study by Mark Changizi showed that wrinkly fingers improved our grip on wet or submerged objects.[1] It was found that picking up marbles in water with fingers that were wrinkled, having been submerged for some time, was much more efficient compared with picking up marbles where the fingers had not previously been under water. This works in a similar fashion to the treads on a tyre in the rain. In dry conditions a smooth surface on the skin gives much better adhesion, as seen in smooth racing car tyres which give far better traction. Wrinkling of the skin is caused by constriction of blood vessels near the skin surface, mediated by the nerves, but why does this only happen on the fingers and toes and not elsewhere in the body?

It is likely that this phenomenon probably evolved as a mutation in early semiaquatic hominins to improve their grip when they were harvesting aquatic plants, mussels, molluscs and other slippery aquatic and marine food. It improved hunting efficiency and survival, subsequently becoming an established inherited trait. A similar feature is also seen in macaques[2] who are regular swimmers and feed from aquatic plants, but no other primates have been studied.

Other interesting structures in the dermis of the skin are the Pacinian corpuscles, which are nerve endings in the skin responsible for sensitivity to vibration and pressure. They respond only to sudden disturbances and are especially sensitive to vibration. Each corpuscle is an egg-shaped structure consisting of many concentric layers of tissue. Embedded within this structure is a free nerve ending. When the corpuscle is deformed by pressure, the nerve ending is stimulated (Figure 7.1).

Beneath the dermis lies the hypodermis, which is composed mainly of loose connective and fatty tissues. It is often referred to as subcutaneous tissue, although this is a less precise and anatomically inaccurate term. The hypodermis contains loosely arranged elastic fibres and fibrous bands anchoring the skin to the deep fascia. In humans and in some mammals, such as whales, the cetacea and hibernating mammals, the hypodermis contains fat stores which form an important insulating layer of blubber and/or food store.

There are subtle differences in human skin and hair, however, which may have important evolutionary significance. The hair on the scalp is more rigid, thicker and denser than elsewhere on the body and, if not trimmed or cut, will continue to grow indefinitely, although not necessarily to the length that Rapunzel managed in the well-known fairy tale by the Brothers Grimm. In males, over the lower facial area, a beard may grow as thick as scalp hair but does not grow to such an extent.

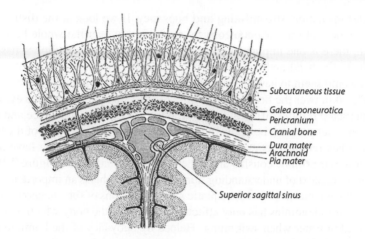

FIGURE 7.2 Human scalp skin. (Courtesy of Henry Vandyke Carter. Henry Gray (1918,) *Anatomy of the Human Body.* (See "Book" section below.) Bartleby.com: Gray's Anatomy, Plate 1196.)

Over the scalp the dermis is densely packed with hair follicles, arrector pili muscles, sebaceous and sweat glands. Between the skin and the underlying muscle is a thin layer of loose connective tissue, which allows the skin to separate easily from the muscle underneath (Figure 7.2). Traumatic injury to the head area may thus result in scalping where the whole hair-bearing area of the scalp may be easily avulsed from the skull, which is not possible in any other part of the body where the skin is closely adherent to the underlying fat and muscle.

Elsewhere in the body, human hair is much finer and is maintained at quite a short length with a thick dermis and underlying fatty layer, which varies enormously in extent and thickness. The dermis is thinner than on the scalp and has a less dense appearance (Figure 7.1). In the groin and under the arms, the hair is moderately thicker and longer, but still maintains a limited length. Some males are more hirsute than others, but females have a much smoother and less hairy skin.

SUBCUTANEOUS FAT

The layer of subcutaneous fat under our skin is another unique feature that distinguishes us from other terrestrial mammals. This adipose tissue is analogous to blubber, which is normal in aquatic and semiaquatic mammals in whom it provides insulation, streamlining and buoyancy. If we look at the distribution and thickness of this subcutaneous fat more closely over the whole body of a human, there is one important exception where this layer of adipose tissue is absent. In the head and neck area there are differences in the skin structure which would seem to have an important evolutionary significance.

The skin over the bridge of the nose, on the forehead above the eyes and over the scalp is quite different from the skin over the rest of the face and neck and over other parts of the body. In these areas over the upper part of the head, the skin has retained its normal mammalian structure. Here, we have a pelt similar to our primate cousins and other terrestrial mammals (Figure 7.2).

In the context of understanding human evolution, this is an important observation. What it means is that the aquatic skin adaptations of subcutaneous fat and loss of hair in hominins has only affected the areas of the body which would be immersed in water when swimming. Helped by buoyancy of the hominin skull from expanded sinuses, the upper part of the face, including the nasal opening for breathing, and the scalp, would necessarily need to be out of the water most of the time. The elongated nasal skeleton and nasal valve in humans would appear to be aquatic adaptations to protect the opening of the vital airway. The eyes would also be out of the water, although when diving we have evolved sight adaptations to help underwater vision. Our ears would be exposed to water immersion, but the delicate eardrum is protected by formation of exostoses.

There does not seem to be any logical reason why any of these unique adaptations in humans could possibly have evolved on the savannah in one single branch of the primate family without even one of these features being present in any of our ape cousins, nor indeed in any of the other savannah or forest mammals who have all remained essentially unchanged over 20 million years. A rational explanation for these human sensory and respiratory adaptations in the upper aero-digestive tract would seem to be much more consistent, however, with evolutionary adjustments to a semiaquatic environment.

EVOLUTIONARY EVIDENCE FOR CHANGES IN MAMMALIAN SKIN

The major changes in skeletal structure in early hominins are evident in the increasing number of fossil remains which have been found. Of the higher primates, the last common ape-human hominin split from our closest cousins, the chimpanzees and the bonobos, occurred about 6 million years before present (mybp). Fossil evidence has shown that the earliest bipedal hominin humans appeared about 4.4 mybp (*Ardipithecus ramidus*) and it seems that they were also tree climbers. We presume that they all had a fur coat similar to extant primates such as the apes and chimpanzees. We know a lot more about bipedal *Australopithecus* from 3 to 4 million years ago and many of the later branches of the hominin family including *Homo habilis*, *Homo heidelbergensis* and *Homo ergaster*, described in earlier chapters.

Soft tissue changes, however, such as skin structure, hair loss, development of subcutaneous fat and changes in thermoregulation are much more difficult to date accurately because of the lack of direct fossil evidence. We can only estimate the timing of these evolutionary changes indirectly by looking at other scientific data. Finding fossil evidence of ear exostoses from the time of *Homo erectus* and Neanderthals confirms, however, that a semiaquatic lifestyle had already evolved by 1–2 million years ago.

Numerous theories have been put forward to explain these evolutionary soft tissue adaptations since the time of Darwin. He wondered why we had so little hair and commented in *Descent of Man* that 'No one supposes that the nakedness of the skin is any direct advantage to man; his body, therefore, cannot have been divested of hair through natural selection'. He thought that we lost our fur coat for reasons of sexual selection and that hairless individuals were more attractive.[3]

However, one of the main problems with most of the theories which have been proposed over the past 150 years about why we lost our hair are based on the savannah ape hypothesis, which suggested that early hominids left the trees for the open grassland where they stood upright and started hunting on the savannah. There they would have been subjected to an open exposed habitat with marked fluctuations in temperature during 24 hours. Their requirements for heat control and preservation would have been quite different compared to those living in a waterside habitat.

As previously explained, recent scientific evidence has convincingly shown that the semiaquatic lifestyle was a much more credible scenario for early hominin evolution rather than the savannah habitat. In 1974, the Russian scientist Vladimir Sokolov wrote a major work on the comparative anatomy of the skin of around 500 mammalian species, in which he observed that 'in water fur provides poor insulation and becomes atrophied'.[4]

Large mammals, like whales, dolphins and other cetacea who live a totally aquatic existence, have lost their fur and have replaced it with a layer of subcutaneous fat, or blubber, which provides insulation, buoyancy and streamlining (Figure 7.3). There are only three clades of mammals that store fat sub-dermally (as opposed to intramuscularly). These are cetaceans (whales and dolphins), pinnipeds (seals, penguins, sea lions and walruses) and humans. Pigs and bears also store fat sub-dermally, but all are either semiaquatic or they are hibernators.

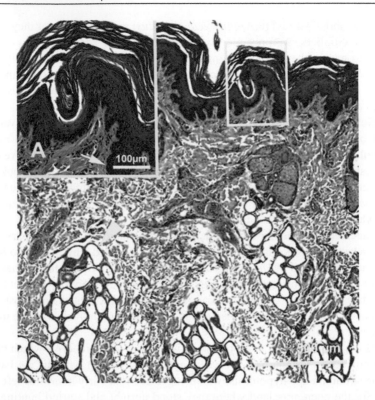

FIGURE 7.3   Seal skin showing deposits of fat/blubber in subdermis. (Courtesy of Khamas, Wael and Smodlaka, Hrvoje and Leach-Robinson, Jessica and Palmer, Lauren. (2012). "Skin histology and its role in heat dissipation in three pinniped species." *Acta Veterinaria Scandinavica.* 54. 46. 10.1186/1751-0147-54-46.)

Other large seagoing mammals, such as the seals and sea lions who go ashore to breed and are therefore semiaquatic have retained their fur because they live in quite cold climates, but have supplemented this, like cetaceans, with a layer of blubber, which is the best insulator in water.[5] Fur seals and sea otters have a combination of a very dense inner air-containing, insulating layer and an outer longer, greasy, streamlined coat. The second largest seal, the walrus is naked apart from his splendid moustache, and the largest species, the elephant seals, have to spend about 40 days each year completing their annual moult. The smallest aquatic mammals such as the water shrew depend on an oily fur coat which can retain a layer of insulating air around the body.

Other large land mammals such as the rhinoceros, hippopotamus and the elephant, whose closest cousin is the sea cow, are somewhat aquatic. They have lost their hairy coat because they live in subtropical climates, although they do use mud to protect their naked skin in the heat of the day, as do pigs who have also lost their hair. This is in contrast to the extinct woolly mammoth which lived in colder parts of Europe and Asia and needed to retain its protective thick fur coat for heat preservation.

Mammals are either terrestrial (horses, cats, lions, apes and dogs), aquatic (cetacea like whales and dolphins) or semiaquatic (seals, beavers, stoats, hippopotamus, etc.). There are only six clades of mostly hairless mammals: cetaceans (whales and dolphins), hippopotimidae, rhinoceridae, elephantidae, heterocephalus (naked mole rat) and humans. One is purely aquatic; three are enormous, subtropical and semiaquatic; one is subterranean; and then there are us humans.

Terrestrial mammals have a hair/fur pelt with very little subcutaneous fat. Even in tropical countries with temperature fluctuations over 24 hours, terrestrial mammals need their hair/fur coat to keep warm at night. Semiaquatic mammals (beaver, stoat, seal, polar bear), have retained their hair because they too need to keep warm at night, living and breeding on the land, but their hair coat is streamlined and waxed for swimming and diving. Some large semiaquatic mammals in hot climates (hippopotamus) have thinner hair coats because they are in the water most of the time, where the temperature remains fairly constant, day and night. Apart from aquatic or semi-aquatic mammals, the only other mammal to be naked, with the exception of humans, is the subterranean Somalian naked mole-rat.

So where do humans fit in? Are they terrestrial, or do they have more characteristics of a semi-aquatic mammal? The evidence suggests that humans have evolved from the quadruped ape over a few million years within this spectrum of semi-aquatic mammals, acquiring a streamlined bipedal shape with strong legs, ideal for swimming, diving and running; a layer of subcutaneous fat to help with insulation; and streamlining and loss of body hair to reduce surface drag and resistance when swimming and running. This seems to be an ideal and efficient combination for a semi-aquatic, waterside ape, which helped to secure the dominant position of *Homo sapiens* 50–100,000 years ago in the harsh and changing environment in Europe towards the end of the last ice age.

Why We Lost Our Coats

The Early Hominin Tailors

The aquatic hypothesis is an ingenious theory which does explain why we are so nimble in the water today, while our closest living relatives, the chimpanzees, are so helpless and quickly drown. It explains our streamlined bodies and even our vertical posture and also clears up a strange feature of our body hair tracts.[1]

Desmond Morris
The Naked Ape

With the exception of modern humans who inhabit every continent, most primates live in tropical or subtropical regions of Africa, America and Asia. So why did humans, who lived in temperate climates with considerable fluctuations in ambient temperatures, either at night or due to seasonal variations, dispense with their protective fur coat, unlike any other terrestrial mammal? We know from genomic skin lice studies that this probably occurred around 100,000–140,000 years ago, at a time when there were several hominin species, including *Homo sapiens*, in Africa; Neanderthals in Europe at the height of the last ice age; and *Homo erectus* in various parts of the earth.

It would seem that, unlike any other terrestrial or semi-aquatic mammal, early hominins were able to dispense with this protective hair covering for three possible reasons:

1. *Increased intelligence*: By about 1.8 million years before present (mybp) the cognitive skills of *Homo erectus* were developing fast. These skills were associated with a larger brain which by then

73

was quite similar in size to anatomically modern humans (AMHs). Living in a semiaquatic habitat gave *Homo erectus* a diet rich in docosahexaenoic acid (DHA) and arachidonic acid, providing the two lipoproteins vital for increased cerebral development and bigger brains.

2. *Fire*: Unlike any other terrestrial or semi-aquatic mammal, by 1–1.5 mya, *Homo erectus* had learnt about fire which had several vital advantages. First, cooking food allowed easier chewing and digestion, and cooking otherwise inedible or tough meat, fish and vegetables provided extra calories available to fuel energy for the brain and body. Second, it gave protection from predators, particularly at night, so that they were able to live and camp on the ground and in caves rather than in trees. Third, it gave warmth to help maintain body temperature. Cooking their food also meant that they no longer had to spend 4–5 hours each day chewing tough, indigestible food like chimpanzees and other primates.

3. *Manual dexterity*: The fact that humans had been bipedal for several million years meant that they had been able to evolve increasing manual dexterity. From fossil evidence we know that *Homo habilis* started using tools around 2.6–1.5 mya in Africa. By 0.5–1 mya, early hominins were no longer dependent on a permanent hair coat because they had developed sufficient skill and manual dexterity to make their own from animal skins.

These primitive hominin tailors were able to adapt their new clothing to suit their ever-changing needs, depending on their activity and the variations in weather and season. They no longer had to put up with a fixed permanent hairy coat like other apes with restricted variations in thickness and limited adaptability to fluctuations in temperature. Instead, the ability to adjust the thickness of their coat and heat regulation within a few minutes gave them enormous freedom to adapt their activities to their environment. When they were cold at night or during the winter months, they could easily adjust their clothing to maintain and regulate their body temperature. This was particularly important when they were on the move, away from the warmth of the camp fire. During the day when it was warm, or when they were hunting on land or swimming and diving for food, they could dispense with their coat to improve their agility and movement.

DESMOND MORRIS AND HOMININ'S 'CHRISTENING CEREMONY'

In his book *The Naked Ape*, published in 1967, Desmond Morris explores different theories about why humans have lost their hair. He mentions the aquatic theory briefly as 'an ingenious theory' which does explain 'why we are so nimble in the water today, while our closest living relatives, the chimpanzees, are so helpless and quickly drown'. He acknowledges that 'it explains our streamlined bodies and even our vertical posture'. It also 'clears up a strange feature of our body hair tracts. Close examination reveals that on our backs the direction of our tiny remnant hairs differs strikingly from those of other apes. In us they point diagonally backwards and inwards towards the spine. This follows the direction of the flow of water passing over a swimming body and indicates that, if the coat of hair was modified before it was lost, then it was modified in exactly the right way to reduce resistance when swimming'.[1]

The finding of aural exostoses in early hominin skulls provides fossil evidence that *Homo erectus* and Neanderthals were swimming and diving on a regular daily basis, hunting for food.[2] It would seem that our hair tracts were gradually modified, as Desmond Morris explains, to facilitate swimming by reducing water resistance. After a period of time, probably hundreds of thousands of years, however, living in a semi-aquatic environment with daily exposure to water, the hair gradually thinned and became atrophied, as observed by Sokolov.[3] Evidence from lice genomics, as mentioned above, indicates that the loss of body fur may be as recent as 100,000–140,000 years ago.

Desmond Morris goes on to explain that we are 'unique among primates in being the only one to possess a thick layer of subcutaneous fat, as an equivalent to the blubber of a whale or seal, as a compensatory insulating device'. Morris stresses that 'no other explanation has been given for this feature of our human anatomy'. He also mentions about 'the sensitive nature of our hands' possibly 'acquired to feel for food in the water' which then was advantageous to the hunting ape. Writing at a time when the savannah theory was accepted dogma to explain nakedness, Morris claimed that, unlike more specialized carnivores such as lions and jackals, the ex-vegetarian ancestral ape was not physically equipped to 'make lightening dashes after his prey' and would 'experience considerable overheating during the hunt, and the loss of body hair would be of great value for the supreme moments of the chase'.[1]

Apart from very hirsute males, hair distribution in humans is restricted to the head, under the arms and in the groin area. It is only over the scalp that the hair is thick, which is thought to provide protection to avoid overheating of the brain from direct sunlight. Brain tissue is extremely sensitive to heat and must be kept within a narrow temperature range to avoid cell death.[4] Mammals which are exposed to the tropical sun in the middle of the day have evolved a number of strategies to reduce overheating of the brain, including cooling of cerebral blood by panting in baboons,[5] a complex nasal heat exchange mechanism in some antellopes,[6] or just seeking dense shade in the middle of the day.

For early bipedal hominins living and foraging on the savannah and open equatorial environments, Professor Peter Wheeler, dean of science at Liverpool John Moore's University, proposed an idea to prevent overheating of the brain, which has been widely accepted. He used a physiological model to demonstrate that hair has been retained in those skin surfaces which are most directly exposed to the incoming radiation over the scalp, the shoulders and upper arms.[7] Our great ape cousins and other primates, on the other hand, who have remained quadruped, have needed to keep their thick coat on all surfaces.

In 2011, Ruxton and Wilkinson pointed out that a hominin walking and hunting on the open savannah would generate a great deal of endogenous heat, in addition to the exogenous radiation from the sun, resulting in a substantial heat load. They concluded that bipedalism must have evolved for some other reason.[8,9] This supports the waterside theory, in which the cooling effect of water immersion would help to dissipate the extra heat load from walking and hunting on land. We can all recognize the desire to cool off in the water on a warm day, especially if we have been exerting ourselves. Several semi-aquatic mammals such as the pig, rhinoceros, elephant and hippopotamus need to wallow in order to keep themselves cool on a hot day, and some cattle and domestic dogs might lie down in puddles. This characteristic human behaviour, however, is not seen in apes or in many other terrestrial mammals.

Desmond Morris's final comment about the aquatic ape theory was that 'despite its most appealing indirect evidence, the aquatic theory lacks solid support' which at that time, 50 years ago, was not available. 'Even if eventually it does turn out to be true, it will not clash seriously with the general picture of the hunting ape's evolution out of a ground ape. It will simply mean that the ground ape went through a rather salutary christening ceremony'.[1]

However, we now do have that 'solid support', previously lacking, in the form of aural exostoses, which could be recognized as the fossil 'missing link' between ape and man, confirming hominin's early 'christening ceremony'.

Morris's comments about the aquatic ape inspired Elaine Morgan to write her first book *The Descent of Woman* in 1972, the first of her four best-selling books on the aquatic ape theory. She raised serious questions about traditionally held beliefs about early hominin evolution which centred around the 'Tarzan-like figure of the pre-hominin male who came down from the trees, saw a grassland teeming with game, picked up a weapon and became a Mighty Hunter'.[10] This rather stereotyped version of supposed events has been taken literally by many authors who have tried to explain hominin nakedness in relation to the savannah theory. Ingenious solutions have been proposed to explain loss of hair and the question of when early hominins first started to wear clothes.

One suggestion relates to the question of lice. Humans are host to three different types of lice: the body louse, the pubic louse and the head louse. Dr. Stoneking and colleagues from Leipzig in Germany studied the DNA of the human head and body lice from around the world and, for comparison, the lice from chimpanzees.[10] They found that the human body louse, which unlike other lice that infect mammals, clings to clothing, not hair, and evolved fairly recently, between 72,000 to 42,000 years ago, and that early hominins would have been wearing clothing since that time. They suggest that this might have coincided with the period when *Homo sapiens* made the exodus from subtropical Africa into southern Europe during the last ice age and may have been an important factor in their subsequent successful domination over the Neanderthals because they were better adapted to cope with the cold climate. It was also a time when their other sophisticated developments of advanced tools, art and trade were rapidly evolving.

On the other hand, Dr. Klein, the Stanford archaeologist, suggested that the Neanderthals and other hominins such as *Homo erectus* must have worn clothing of some kind to have survived in the colder climates of Europe and the Far East.[11] Perhaps the clothing that they wore was more loose-fitting or of different material that was not attractive to human body lice.[12]

SKIN COLOUR, ULTRAVIOLET IRRADIATION AND CUTANEOUS CANCER

Dr. Alan Rogers, an evolutionary geneticist from the University of Utah, has studied an indirect method to try and determine when humans lost their fur by analysing a gene that determines skin colour.[13] The MC1R gene specifies a protein which serves as a switch between two kinds of pigment made by human skin cells. Eumelanin, which protects against the ultraviolet (UV) light from the sun, is brown-black, and pheomelanin, which is non-protective, is a red-yellow colour. In 2000, Dr. Rosalind Harding at Oxford University and others found that the protein made by the MC1R gene tended to vary a great deal outside Africa, but was invariably the same in the African population. She concluded that for survival, it was essential that in African populations the Eumelanin gene remained constant to give protection against ultraviolet light and that there was a powerful genetic survival drive to keep the dark pigment and prevent mutations. In populations in more temperate climates, mutations readily occurred, with variation in the protein, producing skin colours that were variable.[14] As such, these variable skins colours did not give the same protection against UV radiation from the sun.

The colour of human skin in different regions of the world largely depends on the amount of radiation that reaches that part of the planet and is related to the proportion of eumelanin and pheomelanin in the sub-dermis. Our closest primate relatives, the chimpanzees, have dark, protective thick fur and pale skin, without any Eumelanin. It is also very likely that the ancestors that we shared with our primate cousins also had pale skin. The thick hair coat provides vital skin protection from ultraviolet light. However, at a time when all early hominins lived in Africa, their skins were darkly pigmented. So how did the early *Australopithecines* evolve from a dark hair-covered pale skin to a hairless darkly pigmented skin?

There have been many competing theories, but, as with other evolutionary changes, the essential motivating factor is survival and passing genes and any beneficial mutations on to the next generation. Long before the second exodus out of Africa around 100,000 years ago, when early hominins were in East Africa, there must have been some survival advantage for dark skin over pale skin (of ancestral chimpanzees) and this may well have happened when hominins lost their dark protective fur coat because of hair atrophy as a result of their semi-aquatic lifestyle.[15]

The exposed pale skin would have been a liability for several reasons, and any genetic mutation leading to formation of the protective darkly pigmented protein eumelanin would confer a survival advantage to hominins living in sub-equatorial Africa. From deoxy-ribonucleic acid (DNA) analysis, recent evidence has confirmed that the early inhabitants of Britain, the Cheddar Gorge Man, dating from around 10,000 years ago, had dark skin and blue eyes. After settling in Britain and presumably elsewhere in Europe and America with cooler climates, melanin mutations readily occurred from eumelanin to pheomelanin, resulting in paler skins. The DNA from this early Cheddar Gorge Man now comprises roughly 10% of the genome of most white inhabitants living in the United Kingdom.

One possible explanation for the change from pale to dark skin is that the dangerous ultraviolet radiation destroys folate – a molecule which is essential for a wide variety of body processes. For example, folate deficiency causes anaemia; during pregnancy, a lack of folate can contribute to birth defects such as spina bifida. Darkly pigmented skin would therefore protect folate from being broken down. Once our ancestors had lost their protective hair coat, those who did not produce brown-black eumelanin protein in the skin may have suffered folate deficiency which meant that they might not be as healthy to reproduce, or they may have produced fewer and less healthy offspring. Others who had the MC1R genetic mutation to produce eumelanin would have the protective pigmentation and would have left behind more healthy offspring and would have survived more successfully. Over many generations their MC1R genes would have spread through our ancestral population.[15]

Numerous mechanisms have evolved in human skin to sense environmental stimuli and to initiate adaptive responses in order to maintain integrity of the skin and to protect the body from potential harm. Kostyuk and colleagues identified squalene as a probable skin surface lipid (SSL) mediator which reacts with UV irradiation and initiates a metabolic and inflammatory response in human keratinocytes (skin cells).[16] These lipids play an important part in essential skin functions such as mechanical and chemical barriers, thermoregulation and photo-protection. Lying on the surface of the skin, SSLs are exposed to the highest doses of UV irradiation and they form the first line of defence against its potential danger.

Another UV-sensitive receptor is the epidermal growth factor receptor (EGFR), located on the cellular membrane of keratinocytes, which is a key regulator of numerous essential processes underlying skin development, stress responses and repair. EGFRs stimulate keratinocytes to multiply while simultaneously stopping them from becoming more specialised. The overall effect of mutations encourages tumours such as skin cancer to develop. Research reveals that EGFR normally protects these skin functions by preventing expression of a gene essential for tumour development.[17]

UV light plays a fundamental role as an initiator and promoter of carcinogenesis of skin squamous cell carcinoma (SCC), allowing the accumulation of genetic alterations and mutations that gives a selective growth advantage. Normally, the extracellular signals determine whether the cells move from a quiescent state into an active proliferative state. In tumour cells, an increase in the production of growth factors and the growth factor receptors can be often seen to facilitate cellular division.

SCC of the skin ranks second to basal cell carcinoma (BCC) in the frequency of all cutaneous tumours. Its incidence has risen significantly due to an increase in sun exposure and the number of immune-compromised patients. It has a well-defined progression with known precursor lesions called actinic keratosis. The degree of cellular differentiation, tumour thickness, location, and other features have prognostic value. It has a better prognosis than mucosal SCC of the head and neck, also called head and neck squamous cell carcinoma (HNSCC).

The squamous cell lining of the tongue, mouth, tonsil, throat and larynx is clearly not exposed to UV light, but there are other harmful carcinogenic stimuli such as tobacco and alcohol which cause mucosal HNSCC. Human papilloma virus (HPV) causes cervical cancer and is also the major cause of oropharyngeal cancer, which is the most rapidly developing cancer in young and middle-aged adults in the western world. Although the incidence of some HNSCC has decreased because of reduction in smoking, HPV-positive SCC of the tonsil has doubled in incidence in the United Kingdom, northern Europe and America in the last two decades. This is partly because far fewer children have their tonsils removed, but mainly because of changes in sexual behaviour, an increase in sexual promiscuity and number of partners, and earlier age of initial sexual encounter.[18–20] HPV vaccination has been offered to young teenage girls in many countries to help prevent cervical cancer, and in some countries to boys as well. But it was only in July 2018 that the vaccination was available also to boys in the United Kingdom on the National Health Service (NHS) to prevent a further alarming increase in mouth and throat cancers.

Drugs targeting the EGFR are currently the only significant non-surgical option for advanced SCC beyond radiotherapy and conventional chemotherapy. Targeted molecular therapies are becoming increasingly widespread, and an understanding of the evidence for their use as well as their side effect profile is important in order to offer patients informed and current advice.[21–23]

Another possible theory which was previously discounted is the association of ultraviolet radiation damage and the risk of skin melanoma. Research has shown that African Americans are about one tenth as likely to develop malignant melanoma of the skin and less than 1/50th as likely to have a squamous or basal cell carcinoma of the skin compared to white Americans. Previously this was not thought relevant because the majority of skin cancers develop some years after the end of their reproductive lives. If a disease strikes after the person has produced children, a protective gene may not be passed on to the next generation.

However, a new study on skin cancer rates in albino Africans living in central Africa, in the same area where our early ancestors first evolved pigmented skin, has suggested otherwise. Most people with albinism do not have the gene to produce eumelanin and are therefore much more at risk to develop skin cancers than do normally pigmented individuals. In fact, the risk is so high that many of these patients developed life-threatening skin cancers before or during their reproductive years. The data suggests that more than 90% of albino individuals living in this part of Africa will die in their thirties or earlier, mainly because of skin cancer. In one series of more than 500 albino Tanzanians, nearly all died of skin cancer before the age of 40.[24]

Some scientists believe that there are a lot of similarities between our pale-skinned hominin ancestors evolving in Africa and modern-day albinos living in that part of Africa. Perhaps it was a combination of the two adverse factors of folate deficiency and a susceptibility to life-threatening skin cancer that genetically directed the evolutionary pathway towards a dark pigmented skin in our ancestors who were adapting to their semi-aquatic lifestyle where it had been beneficial for them to lose their protective hair coat. The skin cancers in albinos are particularly aggressive and certainly in the days of our pale-skinned ancestors, who did not have the benefit of protective clothing, shelter, high sun factor skin creams or other medical treatments, these cancers may have been much more devastating in eradicating those individuals who did not possess the vital protective genetic sequences in their hereditary chromosomes to pass on to future generations.

THE SIGNIFICANCE OF HOMINID'S 'CHRISTENING CEREMONY'

The significance of our uniquely naked skin compared with other primates and terrestrial mammals is most logically explained by our evolution from our ape cousins as a semi-aquatic rather that a terrestrial mammal. Scientific evidence has convincingly shown that the many anatomical and physiological differences in our skin have gradually evolved over several million years in early hominins as a direct result of our change to a waterside habitat.

For simple chimpanzees venturing into the marshes, lakes and rivers in search of new sources of food, the chimpanzees would become more used to an upright gait in the water over time, allowing gradual rotation of the pelvis and strengthening of the lower limbs. It is suggested that there were very gradual changes in the skin to accommodate this semi-aquatic lifestyle, particularly when the chimpanzees started swimming and diving for a wider variety of food on the river/sea/lake bed. These would have evolved through physiological or anatomical adaptations, natural selection or from genetic mutations which conferred survival advantage.

1. The direction of the hair tracts on their backs changed to point more diagonally inwards towards the spine to reduce resistance when swimming.
2. The hair gradually thinned and became atrophied, as observed by Sokolov.[14]
3. A layer of subcutaneous fat developed to aid buoyancy, insulation and streamlining.
4. The thermo-regulatory changes in the skin gradually evolved with loss of apocrine glands and proliferation of eccrine glands.

5. From a chimpanzee with pale skin and a thick fur coat, the loss of hair would have exposed the pale skin which would be much more sensitive to the dangerous U/V irradiation. The mutation of the MC1R gene which produced the protective dark-staining eumelanin would have been very beneficial, producing a survival advantage for the dark-skinned hominin to prevent damage from the sun and minimizing the risk of developing skin cancers and destruction of skin folate.

From recent genetic evidence from the Cheddar Gorge Man, it would seem that the darkly pigmented skin was still present in some early hominins who came to Britain, but as Harding pointed out,[14] once populations moved away from the tropics, they no longer needed the protective eumelanin, and genetic mutations readily occurred to the paler pheomelanin which produced skin colours that were variable.

We are now realizing the dangers of UV irradiation, which produces skin cancers and melanomas, particularly as there is now freedom of travel all over the globe and many modern *Homo sapiens* enjoy the physical and psychological benefits of sun exposure. Populations who have lived for thousands of years in Europe, North America, northern Asia and other temperate climates have lost the protective eumelanin MC1R gene, resulting in paler skins and the greatly increased risks of skin cancer and melanoma. This is especially evident in Europeans who have migrated to Australia over the last two to three centuries.

Perhaps the desire for some of us to get a good suntan is not only because we think that a darker skin is more attractive, but it may well be a subconscious innate survival instinct to restore a darker protective skin colour. However, it would seem that reversal of the MC1R genetic mutation back from pheomelanin to eumelanin is not so easy to achieve, and waiting for a further natural reversing mutation may take many more hundreds of thousands of years, as it did when the early pale-skinned aquatic hominid chimpanzees began to lose their protective fur coat. Fortunately for these chimpanzees, the eumelanin gene provided protection; meanwhile, we shall just have to be vigilant in using protective clothing and plenty of high sun protection factor sun cream.

Evolutionary Adaptations in the Human Skull and Sinuses

9

Anatomy is to physiology as geography is to
history; it describes the theatre of events.
Jean Fernel (1497–1558)

The French Renaissance physician Jean Fernel studied mathematics, philosophy, astronomy, anatomy and functions of the human body, and he was professor of medicine at the College de Coenouailles for over 20 years. He made many anatomical contributions, including observations about the spinal canal and brain and introduced the concept of physiology. His famous quotation (noted above) provides an eloquent correlation of the two essential sciences which represent the structure and function of the human and animal body. He was one of the first to recognize the way that evolutionary changes in anatomy are closely associated with their functional needs.

In mammals and in other animals, there is nowhere else in the body as important in determining the sensory awareness of the environment as the upper aero-digestive tract, where vital organs of sight, hearing, smell, taste and to some extent touch are situated. They are essential for receiving information about the immediate surroundings for protection, feeding, navigation, communication, socialization and reproduction. Their anatomy and physiology are evolved and adapted to provide the optimum advantage for them to survive and reproduce in the particular habitat or ecological niche in which they exist. The nose is particularly important in detecting changes in the immediate environment.

In hominins, the eyes, ears, nose, sinuses and throat are therefore very important areas of the body which have changed and evolved over the past 5–10 million years from the common ancestral ape to adapt to changing environments and needs. In this and the next few chapters, we shall look at the comparative anatomy and physiological aspects of various structures in the head and neck area which have evolved into what we recognize today as modern humans. Although most of these evolutionary changes have been beneficial for survival, some of them may have had adverse effects on our health. It is fortunate that humans have also evolved a superior intelligence which, unlike anything else in the animal kingdom, has allowed us to recognize, control and treat these medical conditions which otherwise might well have led to our extinction long ago.

The paranasal sinuses are a group of four paired, air-filled spaces that surround the nasal cavity and are in close proximity to the eye sockets (Figure 9.1). The maxillary sinus is the largest of the sinuses and is situated in the cheek area, below the eyes. Above the eye, in the middle of the forehead are the frontal sinuses which do vary considerably in size. Situated between the eye sockets and the nasal cavity are the ethmoid sinuses, which vary in number on each side from 10 to 15. They form a honeycomb-like group of air spaces which are separated from each other by thin bony septa. Finally, the sphenoid sinuses lie at the back of the nasal cavity, behind the orbits.

The functional role of the sinuses has long been in dispute and as yet no satisfactory explanation has been offered for these 'unwanted' spaces. Sinus problems are common ailments from which many humans suffer, but it is surprising

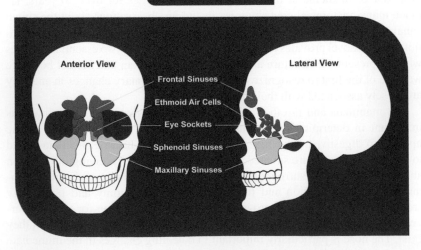

PARANASAL SINUSES

Anterior View   Lateral View

Frontal Sinuses

Ethmoid Air Cells

Eye Sockets

Sphenoid Sinuses

Maxillary Sinuses

FIGURE 9.1   The paranasal sinuses. (Courtesy of udaix/Shutterstock.)

that the only other mammals to have similar problems are breeds like Pekinese dogs and boxers, in which selective breeding has artificially reduced the length and efficiency of the nasal passages.[1] Hay fever and rhinitis, large adenoids and middle ear infections, tonsillitis and sudden infant death syndrome' (SIDS) are also unique problems that affect humans and rarely other terrestrial mammals.

One wonders whether these various medical problems are due to design faults in the upper aero-digestive tract that have been acquired during evolution, since they do not seem to convey any advantage and cause a lot of suffering, particularly in children. When we compare the skulls of the gorilla and *Homo sapiens*, one of the striking differences, apart from the difference in the size of the brain, is the heaviness of the gorilla's skull and the delicate structure of the human facial skeleton (Figure 9.2). This is largely due to the presence of expanded paranasal sinuses. But the answer to the question about why we suffer from these and many other medical problems is that mammals evolved quite happily over 50 million years as a quadruped terrestrial animal and were not designed to have a vertical, bipedal gait, which seems to have caused all these problems.

The Greek physician Galen (130–201 AD) is credited with the original description of skull porosity, but it was Leonardo da Vinci who clearly demonstrated for the first time the existence of the paranasal sinuses in his publication *Two Views of the Skull* around 1489 (Figure 9.3). Direct reference to the sinuses was made later by Vesalius in his treatise *De Humani Corporis Fabrica* published in 1543.[2] However, the functional role of the paranasal sinuses in higher primates, and particularly in humans, has remained a mystery despite

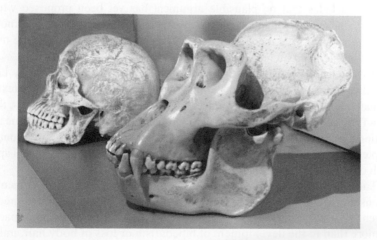

FIGURE 9.2 Comparison of gorilla and human skulls. (Courtesy of Wilburn White/123RF.)

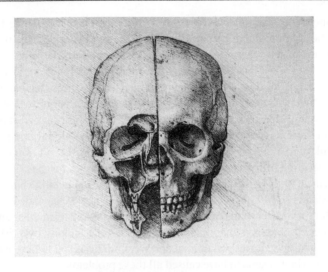

FIGURE 9.3 Leonardo da Vinci's *Two views of the skull.* (Courtesy of Jakub Krechowicz/Shutterstock.)

the formulation of many theories since the time of the Renaissance to explain their anatomical or physiological significance.

The eminent otolaryngologist, Sir Victor Negus, who devoted much of his life to the study of the comparative anatomy and development of the upper aero-digestive tract, discounted these theories but was unable to offer any possible alternative.[3] More recently, in his essay on evolution of the sinuses, Blaney concludes that no plausible argument has yet been proposed which offers a satisfactory explanation for their existence.[4] Takahashi also acknowledges that determining the significance of these air spaces is one of the most difficult problems in human evolution.[5]

Although many of these theories of paranasal sinus function may have seemed superficially plausible at the time of their conception in the context of the anatomical and physiological development of modern *Homo sapiens*, they do not fully address the reality of the evolutionary process in terms of time and functional necessity. Not unreasonably, they are also based on the traditionally accepted savannah theory of evolution of early humans from the arboreal apes, which only recently has been called into question.[1,6,7]

The essential point is that evolutionary adaptations take place in response to things that *have happened*, not things which are predestined to happen.[1] Evolution does not aspire to the development of 'unwanted spaces' any more than to the development of bipedalism or loss of body hair, unless these characteristics provide a definite evolutionary advantage conducive to survival.

That humans possess a large volume of empty paranasal sinus cavities, including an extensive labyrinth of ethmoid cells not found in any other species, must have an important evolutionary significance. There must have been some logical explanation for the expansion of the sinuses in early hominins and development of additional ethmoid cells which offered a definite survival advantage over other ape species.

It can certainly be argued that this legacy left to modern humans has proved much more of a liability than an asset as no other species has the misfortune to suffer from sinusitis or other allied upper respiratory problems. This cannot simply be due to the presence of ethmoid sinuses. Two other contributory characteristics, both also unique to humans, are of equal relevance as aetiological factors.

The first of these is humans' upright posture, which has a distinct disadvantage in providing inadequate drainage of the sinuses and which previously had evolved quite satisfactorily in quadruped mammals over many millions of years. The small apertures, or ostia, which allow drainage of the sinuses into the nasal cavity are no longer situated in the optimum position for the upright bipedal gait for drainage, as they were designed for a horizontally orientated skull. As shown in Figure 9.4, the drainage ostium from the largest of the sinuses, the maxillary sinus, is situated near the roof of the sinus, which predisposes to maxillary sinusitis.

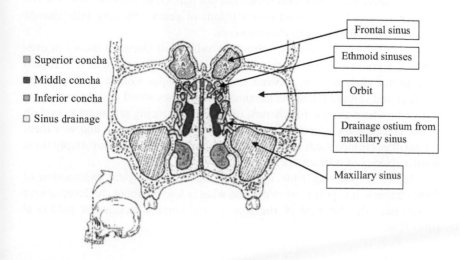

- Superior concha
- Middle concha
- Inferior concha
- Sinus drainage

Frontal sinus

Ethmoid sinuses

Orbit

Drainage ostium from maxillary sinus

Maxillary sinus

FIGURE 9.4 Drainage from the paranasal sinuses.

The second contributory factor is that humans are the only terrestrial mammal not to be an obligatory nose breather. In all land mammals except humans, the larynx is in contact with the soft palate, and the normal airway, from the nasal cavity to the trachea, is thus protected from possible aspiration of liquid and food, which pass around the larynx directly into the pharynx and oesophagus. These mammals are therefore able to keep breathing when eating.

The descent of the human larynx allows the possibility of free mouth breathing and with this comes the unique ability to speak and hold our breath. Despite these obvious advantages, the cost of breathing through our mouths when we are speaking or exerting ourselves, or through necessity because of nasal obstruction from a deviated septum, for example, is significant. With it comes the loss of the important physiological advantages of nasal breathing such as smell, humidification, disinfection and temperature control of inspired air. These factors, predisposing humans to the development of upper respiratory infections, are unique consequences of the various modifications of the upper aero-digestive tract in humans.

Since these modifications to the upper aero-digestive tract in humans seem to have occurred at more or less the same time in their evolution, it is perhaps more relevant to consider them together rather than in isolation. There are also other important features about the human nose, particularly related to its shape and structure, which are quite different from those of all other ape species. These changes could not have evolved for any trivial reason and must represent some significant developmental adaptation that occurred in the vital period of ape/human evolution which did not happen to other members of the ape family, who have evolved over millions of years with very little change into extant modern apes and chimpanzees.

There is certainly no doubt about the validity of Darwin's theory of evolution of humans and apes from a common ancestor. But the widely accepted concept that early humans evolved directly from apes who came down from the trees and ventured onto the savannah, where they stood upright to see further and evolved as hunter-gatherers, offers no convincing argument to explain all the evolutionary adaptations that have occurred in humans, and why there are gorillas, chimpanzees and uniquely different humans and not simply three similar species of African ape.

Before considering what appears to be a more logical explanation of these changes, it is pertinent to discuss what is known about the comparative evolutionary development of the sinuses and current theories of paranasal sinus function.

EVOLUTION AND COMPARATIVE ANATOMY

Study of the evolutionary development of the paranasal sinuses is difficult to consider as a whole since their functions are widely different. The maxillary sinuses are distinct in origin, and while the frontal and sphenoidal air spaces can be grouped together, the ethmoid sinuses are present only in humans, although they are found in rudimentary form in some other higher primates. In general, the size of the frontal and maxillary sinuses was thought to be small in cold climates in order to conserve heat, and Eskimos do have smaller sinuses. On the other hand, Neanderthals, who lived through the last ice age in Europe, have particularly well-developed maxillary and frontal sinuses, which suggests that they had some other function. Certain facts are acknowledged, however, about the functional importance of the paranasal sinuses for respiration, humidification and olfaction in the arboreal ape such as Ramapithecus and in other species during the immediate pre-hominid era (8–15 million years before present (mybp).[8]

MAXILLARY SINUS

The maxillary sinus shows considerable variation in size and is particularly well developed in the higher primates, notably humans, and some ungulates. In chimpanzees, it extends across the floor of the nasal cavity to communicate with its opposite sinus and also with the few ethmoid sinuses present.

To some extent, the size of the maxillary sinus depends on the length of the upper jaw, which evolved primarily for the purpose of seizing and devouring prey. As pointed out by Wood-Jones, the length of the snout varies inversely according to the ability of prehension with the forelimbs.[8] Animals which can grasp food, whether fruit or nuts, as in the case of monkeys or squirrels, or animal prey (e.g. cats), usually have a short snout. Others which have to crop herbage, such as ungulates, some rodents and marsupials (e.g. kangaroos) or predators that can seize prey with their jaws (e.g. dogs, bears and racoons) in general have a long and protruding snout.[3] In many of the carnivorous animals with long snouts, the additional space in the upper jaw is occupied by olfactory maxillo-turbinals (important for smell).

FRONTAL AND SPHENOIDAL SINUSES

The frontal and sphenoidal sinuses are present in many different species but are particularly well developed in keen-scented animals such as the carnivores. Extension of the olfactory mucosa from the nasal cavity into adjacent cavities is seen in members of the cat and dog families, where there is excavation of the sphenoid and frontal bones.[3] Some herbivorous animals also have a large olfactory area in the nasal cavity with specialized mucosa covering five or more ethmo-turbinal bodies.

In both groups of animals, the ethmo-turbinals have a similar arrangement within the nasal cavity, and the remaining space is occupied by the maxillo-turbinals, which are the main source of moisture for humidification of the inspired air, essential for olfaction. The result is that in keen-scented animals, the nasal fossae are filled with ethmo-turbinals and maxillo-turbinals, with little or no free space.

In primates, however, there is a wide difference in the extent of these sinus cavities. The majority have no frontal or sphenoid sinuses, but in humans, these spaces are often well developed without any obvious reason, for their olfactory area is very restricted and they have a feeble sense of smell.[3]

ETHMOID SINUSES

The reduction of the olfactory area in higher primates has resulted in regression of the olfactory turbinals, which are replaced by additional air spaces – the ethmoid sinuses. According to Cave and Haines, chimpanzees have a small anterior ethmoid cell opening into the frontal sinus, and a posterior cell, whereas gorillas have an anterior and two posterior cells.[9]

In humans, however, there is unique development of large and numerous ethmoid sinuses communicating with the nasal fossae by narrow ostia, despite the lack of any olfactory or other apparent role. In addition, the sinus openings are well protected from the nasal airstream by three scroll-like turbinates or conchae which project downwards to cover the ostia, forming narrow valve-like communications (Figure 9.4).

The extensive system of empty ethmoid cavities found only in humans forms a honeycombed labyrinth situated between the upper part of the nasal cavity and the orbit, resulting in characteristic widening of the inter-canthal distance compared with apes (Figure 9.5). There may be as few as three or as many as 18 on each side, their number varying inversely to their individual

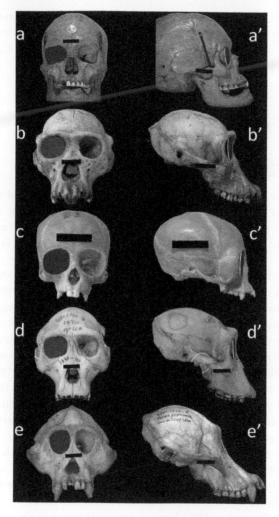

FIGURE 9.5 (a) Modern human, (b) chimpanzee, (c) gibbon, (d) gorilla, (e) orangutan. (3 cm black bar scale on each skull). (Courtesy of Denion, E. et al. "Unique human orbital morphology compared with that of apes." *Sci. Rep.* 5, 11528; doi: 10.1038/srep11528 (2015).)

size. The pyramid-shaped labyrinth has an average length of 4–5 cm, a height of 2.5–3 cm, and a width anteriorly of 0.5 cm widening to 1.5 cm posteriorly, giving an overall volume of up to 30 cc.[10] This results in widening of the nose and also expansion of the frontal area associated with a larger brain.

In the context of higher primate evolution, the emergence of early humans with their short snout but extensive framework of large empty paranasal cavities, despite a rudimentary olfactory or prehensile requirement, poses two apparently unanswered questions: What was the purpose of these empty sinuses? Why did humans develop a profusion of additional ethmoid spaces not present in any other animal?

THEORIES OF PARANASAL SINUS FUNCTION

The hypothetical role of the paranasal sinuses has been the subject of much speculation but little empirical research or investigation over the last few centuries. Several well-known theories have been proposed, but they were later discounted for lack of substantial evidence or conclusive observations.

1. *Resonance theory*: In 1660, Bartholinus[11] originally proposed that the sinuses were important in phonation by aiding resonance of the voice. As Blaney[4] points out, this theory has been discounted because the size of the sinuses bears little relation to the strength of the voice. Animals such as the giraffe and rabbit, despite having large sinuses, have a weak or shrill irresonant voice,[3] whereas others, for example, the lion, although possessing small sinuses, can produce an unmistakable loud roar.[12]

2. *Mucus secretion theory*: In 1763, Haller suggested that the sinuses played an important role in moistening the olfactory mucosa[13] but this theory has been discounted because of a comparative lack of mucus glands in the paranasal sinus lining.[3,14,15]

3. *Olfactory theory*: This theory was first proposed in 1830 by Cloquet, who assumed quite erroneously that the large sinus cavities in humans were lined with olfactory epithelium.[16] As with other animals which have a poor sense of smell, the olfactory mucosa in humans is confined to a limited area in the roof of the nasal cavity and the sinuses are lined with respiratory epithelium.[17]

4. *Thermal insulation theory*: First suggested in 1953 by Proetz, who likened the paranasal sinuses to 'an air-jacket about the nasal fossae closely resembling the water jacket of a combustion engine',[12] this theory has been discredited for several reasons. In species which require the greatest degree of warming of inhaled air for increasing humidification, the heating apparatus takes the form of an elaborately branching maxillo-turbinal body inside the nasal cavity. Heat exchange of the inspired air is much more efficiently carried out with this extensive system of superficial vascular spaces in contact

with the nasal air stream than with large sinus cavities situated adjacent to, but separate from, the nasal fossae. Furthermore, the presence of only one small ostium, or occasionally two, precludes the possibility of adequate circulation of air into the sinuses.[3]

In addition to the humidification theory, it has also been suggested that the paranasal sinuses help with insulation of the base of the brain, but the apparently anomalous presence of large frontal sinuses in the African Negro[18,19] and their frequent absence in Eskimos would seem contradictory to this theory.[20,21]

Neanderthals also possessed large frontal sinuses which according to Coon were adapted to insulate and protect the brain from the cold,[22] but as Tillier points out,[21] this theory cannot be extended to other hominins since the frontal sinus and supraorbital size relationship is unique.[4] Quite apart from this, Neanderthals were mainly accustomed to a temperate climate with little need for partial cerebral insulation.

5. *Lightening the skull*: Another suggested role of the paranasal sinuses is to lighten the skull, particularly its anterior half, in order reduce the work of the neck musculature.[14] Unlike other primates such as gorillas, orangutans, chimpanzees and gibbons, humans are unique in maintaining an upright posture. In these other species the skull is held in a forward inclined position requiring strong neck muscles capable of supporting the head under all conditions (Figure 9.6). Only in humans is the head balanced on occipital condyles situated in the middle of the skull base rather than at its posterior extremity, as in most four-footed animals.[8]

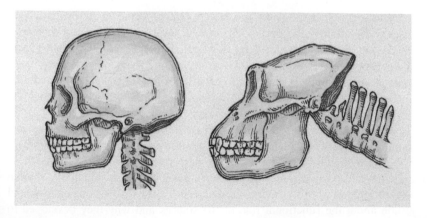

FIGURE 9.6 Cervico-occipital articulation in human (vertical spine directed inferiorly) and gorilla (horizontally directed posteriorly). (Courtesy of Dorling Kindersley/Science Source.)

'If sinuses served no other useful purpose than the supposed reduction in weight', writes Negus[3] 'the obvious alternative would be the apposition of the two tables of bone which form their walls'. He goes on to say that 'this would be a simple matter in the frontal and sphenoid regions' and elimination of the necessity for maxillary sinuses could be evolved, as in baboons, by incurving of the cheeks.

6. *Facial growth theory*: In 1922, Proetz originally considered the presence of nasal sinuses to be directly related to the development of the face and that 'we are not called upon to attribute to these cavities any further functional activity'.[23] He concludes that 'the face parts develop because the individual has need of them: larger, stronger jaws, increased breathing space. The sinuses, which, after all, are nothing but unoccupied spaces, result incidentally'. More recent authors have stressed the importance of cranio-facial development and increase in the angle between the forehead and frontal cranial base in dictating sinus morphology.[4,5,24] But these arguments do not explain the evolutionary necessity for the presence of large empty spaces in the facial skeleton adjacent to the nasal cavity in humans when compared to parallel development in other species.[7] Although this theory may be tenuously applied to the maxillary and ethmoid sinuses, there does not appear to be any reason to regard the presence of the sphenoidal sinus as connected with growth of the face, since it does not form part of the facial structure. Similarly, the wide variation in the size of the frontal sinuses without obvious change in the shape of the facial contour seems to render this argument invalid.

Certain conclusions about the functional importance of the paranasal sinuses may therefore be derived from a general survey of their comparative anatomy and evolutionary development. Expansion of the nasal fossae and adjacent sinuses was almost universally apparent only for the purpose of extension of the olfactory mucosa in carnivores and other keen-scented animals, apart from the higher primates and perhaps one other exception – the excavated sinuses and hollow horns of some ungulates.

Other functions ascribed to the sinuses in humans such as vocal resonance, humidification, heat exchange, thermal insulation and craniofacial development may be discounted for the various reasons described in this chapter. Since these theories, based on traditional evolutionary ideas of the savannah ape/human transition, are not wholly convincing, we are still left with the same unanswered question about the functional role of the sinuses and why they developed to such an extent in humans.[7] If we introduce the concept of aquatic adaptation, however, as proposed by Hardy[6] and Morgan,[1,25,26] many of these puzzling evolutionary changes seem to have a much more logical explanation.

Human Skull Buoyancy and the Diving Reflex

10

> *Modern humans do possess the physiological qualities necessary for making a living from hunting-gathering via breath-hold diving.*
> Abrahamsson and Schagatay (2014)

In the previous chapter, the anatomy and various theories of paranasal sinus function were reviewed. Of these, the theory proposed by Skillern,[1] that expansion of the sinuses in humans was for the purpose of lightening the skull, may be nearer the truth. The counterarguments to this theory pointed out by Negus[2] and others are perfectly valid when taken in context of the traditional savannah theory of ape/human evolution, which was the current understanding of early hominin evolution in the early twentieth century. If simply lightening the anterior part of the skull was needed for reducing the workload of the neck musculature, this could equally be achieved by reduction in sinus size and apposition of their walls.

If we argue, however, that the anterior part of the skull not only needed to be light but also buoyant, the presence of air in the cavities would be essential, and development of larger and more numerous sinuses (e.g., ethmoid cells) would be even more advantageous. From purely a physiological and anatomical point of view, this would seem to be the most logical explanation.[3]

This hypothesis would obviously have no relevance if one continued to accept the traditional savannah theory of evolution. It is consistent with the waterside/aquatic theory, however, which proposes that certain early hominins, ancestral to *Homo sapiens*, spent a period of several million years adapting to a semi-aquatic environment, spending part of their time swimming and diving, hunting for their survival. In their situation, additional buoyancy of the facial structures would certainly be advantageous to aquatic apes in helping to keep the nasal airway opening above water, relieving the workload of the cervical muscles and assisting forward migration of the cervico-occipital articulation.

Other aquatic species have evolved structural buoyancy to suit their individual needs. Those non-air-breathing animals which live an entirely submarine existence, such as fish, require mainly negative buoyancy which is controlled in many species by a swim bladder, but others which need to surface for breathing have additional buoyancy. Aquatic mammals such as the cetaceans (whales and porpoises) have developed a layer of subcutaneous fat (blubber) which provides buoyancy and also streamlining and insulation. This layer of subcutaneous fat is also a unique feature which distinguishes humans from all other primates. If we consider that *Homo sapiens* evolved as a terrestrial mammal, this distribution of fat is highly uncharacteristic, but if *Homo sapiens* evolved as a semi-aquatic mammal, this would be conforming to type.

Maintenance of the airway is obviously essential in air-breathing animals. In whales and other cetaceans, the nostril or blowhole has migrated well on to the dorsum of the head, which allows the snout to remain submerged on the surface while the animal is still breathing. In the reptile family, the long characteristic boat-shaped snout of the crocodile gives additional buoyancy for the airway which is also vital for maintaining its acute sense of smell.

The presence of air-sacs or cavities is one of the most efficient means of assisting buoyancy, as seen in some surface insects. In frogs and toads, the air sacs provide buoyancy for the head while the animal is partly submerged to help keep the nasal opening above water. It is therefore not unreasonable to suggest that in primates, who already have rudimentary sinuses, expansion of the cavities and development of additional ethmoid spaces evolved in early humans for the same purpose.

THE DIVING REFLEX, NITRIC OXIDE AND THE PARANASAL SINUSES

The diving reflex is the name given to an automatic response seen in all vertebrate animals to immersion of the face in water. This reflex overrides the normal haemeostatic controls in order to conserve oxygen and redistribute blood and oxygen stores to vital organs such as the brain and heart. It is especially well developed in aquatic mammals (whales, dolphins and other cetaceans) and semi-aquatic mammals such as seals, otters, muskrats and humans, particularly in babies up to 6 months old. Diving birds such as the penguin also exhibit this reflex, which is thought to be an evolutionary development of a more primitive response to hypoxia in fishes.[4]

Sensory nerves of the face, in particular the anterior ethmoidal nerve, which innervates the anterior part of the nasal cavity, are all branches of the fifth cranial nerve (the trigeminal nerve) and are very sensitive to changes in temperature. When the face is submerged, the nerve immediately relays the information to the brain to stimulate the diving reflex via the vagus nerve (the tenth cranial nerve) which supplies the heart. This causes bradycardia (slowing of the heart rate) and peripheral vasoconstriction, which diverts blood from the limbs and all muscles and organs to the heart and brain. In this way, what oxygen is available in the body goes to the two vital organs.

In humans, the diving reflex is not triggered by immersion of the limbs in water nor by holding one's breath or by water touching the face, but the combination of water submergence of the face and breath-holding induces the maximum response of bradycardia, the strength of the reflex increasing proportionately to decreasing water temperature. The anterior ethmoidal nerve is considered the gatekeeper nerve because it is the first to detect noxious gases or water entering the nasal passages. Division of this nerve eliminates the bradycardia and reduces the apnoea and blood pressure changes associated with the diving reflex[5] which is also known as the trigeminal cardiac reflex (TCR).

There are various mechanisms available to increase oxygen supply to the brain. Recent research by Seymour from Adelaide has shown that, although the hominin brain has increased in size threefold since the *Australopithecines*, its blood supply has increased sixfold[6] (see Chapter 13). The main storage of oxygen is in haemoglobin in the red blood cells and also in myoglobin in muscles. In diving conditions when the muscles become ischaemic and hypoxic from the vasoconstriction, oxygen is released from the myoglobin. In aquatic and also semi-aquatic mammals, investigations have shown that the myoglobin stores in muscle are 10–30 times higher than that found in terrestrial mammals, and myoglobin stores also increase after extended foraging behavior underwater.[7,8]

Another source of extra oxygen is the spleen, which contracts during the diving reflex, releasing a large number of red cells thus making more oxygen available. A recent study by Melissa Llardo from Copenhagen published in the journal *Cell* suggests that a DNA mutation gives rise to enlarged spleens in the Bajau free diving population. It is suggested that a gene called PDE10A is associated with thyroxine production in the Bajau divers and is related to spleen size. These changes are not seen in a similar Indonesian population who live on the mainland. This evolutionary adaptation is similar to that in seals, who are known to have large spleens that provide an additional source of oxygen when diving.[9]

The recent discovery in 1995 of high concentrations of the gas nitric oxide (NO) in the paranasal sinuses by Lundberg and others[10,11] has led to further research into possible important physiological functions of this gas and the significance of its production in the sinuses. It was found that the healthy mucosal lining of the paranasal sinuses produces large quantities of NO, which has important properties of vasodilatation and antimicrobial activity. Hypoxia is one of the most powerful inducers of NO production by increasing nitric oxide synthase activity in the lining of the sinuses.[12] It has been suggested that the diving reflex also activates the production of NO through trigeminal nerve stimulation as a neuro-protective reflex leading to rapid cerebrovascular vasodilatation to help maintain optimum oxygen supply to the brain.[13]

THE NASAL VALVE

One other essential characteristic of air-breathing marine mammals is the ability to close off their airway when submerged. This can be achieved either by a valvular mechanism to close the nostril (newt, crocodile, cetaceans, salamanders, sea elephants, sea lions, seals and polar bears) or the laryngeal aperture (penguins). This provides protection for the airway and allows the animal to catch or eat prey whilst under water. Of tailless amphibians, the frog has a specialized pad on the anterior angle of the lower jaw which is thrust upwards when submerged to close the external naris.[14] The ability to close off the nostrils is not exclusively aquatic, since the camel can also achieve this to keep out sand.

The mechanism of closure of the nostril is by two pairs of opposing muscles, the dilator and constrictor naris, which are situated on the dorsum of the nasal aperture. It is perhaps curious that, although humans are generally unable to completely close off their nostrils, they are unlike any other higher primate in possessing similar muscles around the external nose which can be used to flare the nostrils. The only difference between humans and seals is that when the muscles are relaxed, the seal's nostrils are completely closed, whereas a human's is not.[14]

THE EXTERNAL NOSE

The shape of the human nose has intrigued scientists and anatomists for centuries, but no logical explanation has been offered to account for its elongated form and the unique differences which distinguish it from those of other higher primates. The prominent bony and cartilaginous portion of the nose has made it vulnerable to trauma, causing deviations and fractures. Even during normal childbirth, the prominence of the nose may lead to cartilaginous greenstick fractures of the septum that are not usually apparent during childhood, but during the teenage years when the nose begins to grow, a cartilaginous weakness on one side may lead to gradual deviation of the septum, leading to nasal obstruction.

The protective hooded shape of the external nose does seem to be more suited to swimming and diving by deflecting water from the nasal opening, preventing inundation of the airway (Figure 10.1). In contrast, the other higher primates possess an open anterior-facing nasal aperture, with the exception of the male proboscis monkey, who possesses a very prominent

FIGURE 10.1 The elongated nasal skeleton protects the nasal opening during swimming. (Courtesy of Netfalls Remy Musser/Shutterstock.)

nose. Proboscis monkeys live in the mangrove swamps of Borneo and do spend much of their time wading upright in the water. For the male, his enormous nose is thought to be primarily an attractive male feature rather than designed to protect the nasal airway. The proboscis monkey has been observed walking bipedally on land, as well as in the water, but his natural mode of locomotion is quadruped.

Thompson and Dudley Buxton have studied the configuration of the nose in modern humans in relation to racial origin and climate and found that the wide platyrrhine nose is associated with a hot, moist climate and the narrow leptorhine nose with cold, dry conditions.[15] However, the breadth of the bony pyriform aperture at the end of the nasal skeleton has no correlation with racial differences.[2]

THE FUNCTIONAL ROLE OF THE NOSE AND SINUSES IN A WATERSIDE HABITAT

The extensive labyrinth of thin bony plates covered with epithelial mucosa in the nasal cavities of mammals, known as turbinates or conchae, are very important in vastly increasing the surface area of the nasal mucosa in contact with the inspired and expired air. The turbinates nearer the nasal opening are mainly concerned with respiratory functions of heat and moisture conservation, whereas the more posterior ones have an olfactory role.

In aquatic species, roughly 25% of the nasal cavity area is devoted to olfactory function compared with 70% in terrestrial species; the remaining areas of 75% and 30%, respectively, have respiratory functions. In water, body heat is lost more quickly than on land, and the increased respiratory area in aquatic species has evolved to minimize heat loss. At the same time, there is less reliance on olfaction when foraging under water. Elephant seals, for instance, undergo lengthy fasts of up to 12 weeks during the breeding season, during which they acquire no water. In order to minimize water vapour loss through respiration they use periods of apnoea (cessation of breathing) and also have an extensive area of respiratory turbinates that occupy roughly 90% of their nasal cavity. In this way, they are able to retain 92% of the water of each breath.

In humans the functional role of the paranasal sinuses has long been debated, but their contribution to lightening of the hominin skull in comparison with our primate cousins, together with the presence of new ethmoidal sinuses in humans and the expansion of the maxillary and frontal sinuses in Neanderthals and humans, suggest that they may have been important in providing extra buoyancy in a semi-aquatic habitat. The advanced development of the diving reflex in humans, with all of its physiological changes affecting the heart, respiration, circulation and brain perfusion, would also support the evolutionary adaptations of a waterside ape.

Surfer's Ear

11

*Remnants of the past that don't make sense
in present terms – the useless, the odd, the
peculiar, the incongruous – are signs of history.*
Stephen Jay Gould (1941–2002)

For well over a century, ear, nose and throat (ENT) surgeons have been aware of a curious condition seen in their clinics, in which bony swellings appear to grow in the deep part of the external ear canal. They can see them quite easily when inspecting the ear, but no one has been able to explain exactly what they are and why they occur. The most unusual aspect of these bony swellings lies in the fact that they are only associated with frequent swimming and regular water immersion of the external ear canal and, because they are seen frequently in surfers, they are now known colloquially as surfer's ear. They are never seen in people who do not swim regularly.

Much has been written about the possible cause of these bony swellings, called exostoses, but several aspects still remain unclear. The fact that the condition is always bilateral is fairly predictable since both ears are immersed in water, but why do they only grow in swimmers? Why do they only grow in the deep part of the bony ear canal, just near the eardrum, and not elsewhere in the body in other areas which are exposed to water? Why do they always grow at two or three constant sites? From an evolutionary point of view, what is or was the purpose and function of these rather incongruous protrusions? (See Figure 11.1.)

As we have seen in earlier chapters, palaeoanthropological evidence has modified our ideas about early hominin evolution in recent decades, inspired by Sir Alister Hardy's visionary article[1] and Elaine Morgan's aquatic ape theory.[2,3] This suggests that there was a significant influence in early hominin evolution of an aquatic phase, which resulted in a number of unique features that were not seen in other primates nor in any other terrestrial mammals. Looking at various aspects of comparative anatomy and physiology of the skull, ear, nose and throat area in humans and other primates, there seemed to be several enigmatic anomalies, including these bony ear swellings, unique in humans, that were not seen in other terrestrial mammals. It was difficult to reconcile

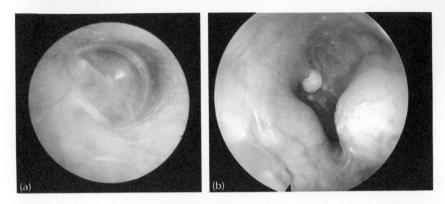

FIGURE 11.1   (a) Deep ear canal: normal eardrum and (b) exostoses of the ear canal. (Left: Courtesy of Clínica Clarós/Science Source. Right: Courtesy of Science Photo Library/Science Source.)

these differences with the accepted assumption that early hominins evolved from the ape family roughly 5–8 million years ago on the savannah. They did seem to make much more sense and were more logically explained, however, by an historical period of aquatic adaptation during early hominin evolution.[4]

I proposed that these exostoses were evolved for the purpose of protection of the delicate eardrum during swimming and diving, by narrowing the ear canal in order to reduce the pressure on the eardrum which might otherwise be prone to rupture.[4] Having studied other mammals who live in a semiaquatic environment, spending part of the time on land but also swimming and diving in search of food, we see similar protective mechanisms when in the water, either partially narrowing or temporarily closing the ear canal when under water. I suggested that if these exostoses were identified in early human fossil skulls, this would provide good evidence to support the aquatic theory in early hominins.[4] In contrast to modern humans, their frequent immersion in water would have been in search of food for their survival. Recent fossil evidence of exostoses found in the external ear canal in early hominin skulls provides the vital evidence of humans' aquatic past.[5,6]

This chapter reviews the current literature regarding external auditory canal exostoses which have been found in recent and ancient archaeological populations. The different adaptations of the mammalian hearing mechanism in aquatic, terrestrial and semi-aquatic mammals are also described. These adaptations show that humans are similar to other semi-aquatic mammals which all have evolved mechanisms to close off or narrow the ear canal when swimming or diving. The developmental embryology of the external ear canal is also described, and it helps to understand why the little bones develop in two or three specific sites in the ear canal, corresponding to the growth plates of the three tiny bones which make up the bony part of the external ear canal.

EAR CANAL BONE ABNORMALITIES

Bony swellings of the external ear canal occur with varying frequency and are of two distinct types: osteomata and external auditory canal exotoses. Osteomata are unilateral, usually arise in the outer part of the canal and have a pedunculated or lobular appearance. Smaller ones are sometimes seen close to the upper part of the eardrum. They are composed of dense ivory bone and are considered to be true pathological benign tumours of bone. There may be a hereditary factor in their aetiology, since Roche[7] has found an incidence of 27.9% in Australian aborigines. Herdlycke[8] has also described a particularly high incidence in Peruvian and American Indian populations.

Exostoses, on the other hand, are bony swellings which arise in the deep part of the external auditory canal. These exostoses are benign, bilateral, broad-based, thickened bony swellings, 70% of which are found on the posterior wall.[9,10] They can also be found situated on the lower anterior wall and sometimes on the roof (Figure 11.1b). Clinically, most are asymptomatic, but they may be associated with otitis externa (infection of the outer ear), and obstructive symptoms may occur when the canal occlusion reaches over 80% (Figure 11.1b). The introduction of high-resolution computerized tomography (CT) has greatly enhanced the ability to demonstrate anatomical bony details.

Even recently, the cause of external auditory canal exostoses and osteomas has been considered an unresolved issue.[11] In the past it was thought to be predominantly genetic,[8] but most researchers now agree that genetics most likely plays a minor role in the development of this trait, and it is now generally accepted that cold water exposure is the main aetiological factor.[12–17] However, the exact way in which water stimulates bony hyperplasia (overgrowth) at these specific sites remains uncertain, but a review of the anatomy and developmental embryology of the external auditory canal may shed light on why water exposure results in hyperplasia in certain sites.

DEVELOPMENTAL EMBRYOLOGY OF THE HUMAN EXTERNAL EAR CANAL

The adult external auditory canal is divided into two portions: the cartilaginous portion in its outer third and bony portion in its inner two thirds (Figure 11.2). It measures 24 mm on average in length. Its development *in utero* is a complex

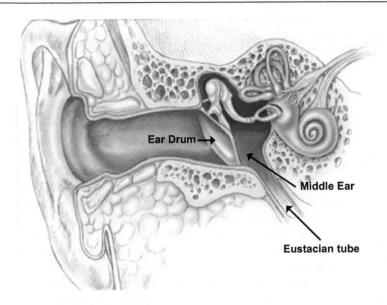

FIGURE 11.2 The human ear showing a normal ear canal. (Courtesy of the author.)

process but could explain the reasons behind the development of external auditory canal exostoses.

The external auditory canal develops from the dorsal (posterior) part of the first branchial groove (analogous to the gill of a fish) during the fourth and fifth week of gestation.[18] From the skin surface, it extends inwards as a funnel-shaped tube which deepens, with proliferation of its ectoderm (skin), forming an epithelial plug.[19] The ectoderm at the deep end of the epithelial plug is briefly in contact with the lining of the pharyngeal pouch (Eustachean tube and middle ear). A membrane then grows between the pharyngeal pouch (middle ear) and branchial groove (external ear canal) and eventually separates the two spaces, forming the tympanic membrane (eardrum).

From 18 weeks gestation, the latter stages of development include opening up and clearing of the deep canal and widening of the whole canal. The bony part of the ear canal begins to ossify during the fourth and fifth month to form the tympanic ring, which supports the tympanic membrane and three other centres of ossification, just lateral to the eardrum: the tympanic, squamous and mastoid growth plates, which grow outwards in a tubular fashion, forming the bony part of the external ear canal (Figure 11.3).[19]

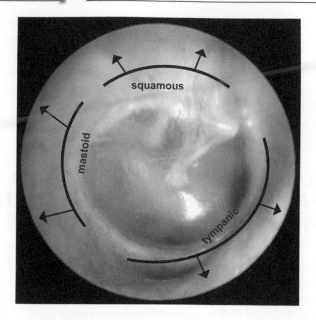

FIGURE 11.3 The three epiphyseal growth plates, just lateral (adjacent) to the margin of the eardrum. (Courtesy of the author.)

In the bony portion of the normal external auditory canal, the distance between the epidermal (skin) surface and underlying bone is small.[20] Pilch believes that the close proximity of the skin to the underlying bone could explain why exostoses develop in this particular area.[21] He suggests that cold water enters the deep canal and exerts a cooling effect that simulates the nearby periosteum (surface layer of the bone) to produce new bone. In light of this, animal studies using guinea pigs have shown that repeated exposure to cold water in the canal causes fibrous proliferation of the subcutaneous tissue in the deep meatus and stimulates the periosteum to produce a layered formation of periosteal bone.[22] Furthermore, Harrison also used guinea pigs in similar experiments and described similar histological evidence of new bone formation in the deep meatus.[23]

The specific location of the three sites of exostoses formation in the external ear canal suggests that these correspond exactly to the three centres of ossification. With repeated exposure to cold water in the ear canal, there is redness and hyperaemia (increased blood flow) at these three ossification sites, and then a resulting cellular inflammatory process initiates a repair response. Similar to a fracture of a long bone, repair must result from the actions of osteoblasts (cells which make bone) and osteocytes (cells which repair bone) which work together to create new bone – exostoses in this case. Long-term exposure of the ear canal to water therefore can result in osteogenic activity[24–26] and formation of exostoses at the growth plates in the deep bony canal wall.

EVOLUTIONARY ADAPTATIONS TO THE HEARING MECHANISM IN AQUATIC AND SEMIAQUATIC MAMMALS

It seems that the formation of external auditory canal exostoses in *Homo* species may represent an adaptive response to an aquatic environment. Land-based mammals which depend on air-mediated sound transmission from the eardrum usually have a widely patent external ear canal (Figure 11.2) for maximum reception of sound. Marine mammals, on the other hand, have adapted to sound transmission in an aquatic medium and have narrow or closed-off canals. A wide ear canal is no longer needed and may also be a dangerous liability, predisposing to rupture of the eardrum because of rapid increase in external pressure when swimming and diving. The evolutionary adaptations to the external ear canal therefore depend on whether the species lives permanently in water or a terrestrial environment or has a semiaquatic habitat.

Whales and other cetaceans have evolved very specialized mechanisms for sound transmission and reception. Since these great mammals left the land and took to the seas millions of years ago, their survival has depended on the evolution of ear structures that are highly specialized for underwater hearing. Because their hearing no longer depended on air-mediated transmission, the external ear canal became completely occluded with a keratin plug and they have developed a highly sophisticated sonar mechanism. They rely primarily on sound for sensory perception, as well as for communication. Sound, therefore, has a profound influence on navigation, feeding, socialization, breeding and other whale behaviour.

Extant whales, or cetaceans, fall into two groups – toothed whales, or odontocetes, and baleen whales, or mysticetes. Toothed whales are fast swimmers and hunters, and include porpoises, dolphins, killer whales and sperm whales. Baleen whales are filter-feeders that live on zooplankton and include the blue whale, the humpback whale, the gray whale and the bowhead whale.

Among the many differences between toothed and baleen whales is their specialization for underwater hearing. The toothed whale's biological

FIGURE 11.4 Toothed whale forehead melon for echolocation. (Courtesy of Miles Away Photography/Shutterstock.)

sonar, also known as echolocation, involves vocalisation from the nose, and the reception of the echo in the ear. The whale recycles the air in its complex nasal passage, and produces a high-pitched sound that is beamed out to the environment through the melon, an oil body in its bulging forehead (Figure 11.4). The sound beamed from the melon into the environment is bounced back to the whale, providing information that helps the animal to accurately decipher its environment through highly specialized modification of the hearing mechanism.[27]

By contrast, the baleen whale can vocalise and hear very low-pitched or infrasonic sound, which can travel great distances and scatter to large areas in water. With infrasonic sound, baleen whales can communicate with each other over geographic areas as large as an ocean basin.

Semi-aquatic mammals, on the other hand, still need air-transmitted sound reception on land but also require protection of the tympanic apparatus when underwater. The hippopotamus can stay underwater for about 15 minutes, and closure of its ear canal is achieved by contraction and angulating the ear canal backwards. The platypus achieves this with a skin fold, and the desman, by a glandular swelling.

Hooded seals show an interesting model of evolutionary adaptation in that they have broad-based exostoses in the floor of the external canal lateral to the tympanic membrane. These exostoses, along with cavernous tissue in the middle ear, allow the seal to dive to depths of greater than 1000 metres at pressures of 100 atmospheres without damage to the middle ear or tympanic membrane.[28] Clearly, the presence of external auditory canal exostoses confers a selective advantage for hooded seals in the marine environment. And so it might have done in hominids.

EXTERNAL EAR CANAL EXOSTOSES IN MODERN POPULATIONS

From early years, the relationship between water exposure and external auditory canal exostoses in modern populations has been described. Two of the most noted otolaryngologists, Belgraver[29] and Van Gilse,[13] found similar populations describing this cold-water relationship. Belgraver, for instance, found that members of a swimming club had an incidence of 42.8% compared to patients in his clinic with an incidence recorded as low as 2.02%. In 1893, Field made the first association between external auditory canal exostoses and sea water.[30] Korner similarly suggested salt water as an important causative factor when he first noted the greater incidence of external auditory canal exostoses in coastal populations compared to inland populations in Germany.[31]

The natural history of external auditory canal exostoses has been observed by Wong and colleagues[32] in a population of 307 surfers. They found that the length of time that individuals spent surfing was proportional to the presence and severity of external auditory canal exostoses. Specifically, they observed that in surfers with external auditory canal exostoses, 61.1% had surfed for 10 years or less, but in surfers who surfed more than 10 years, 82.4% had severe external auditory canal exostoses. Mann further noticed in his study that severity of external auditory canal exostoses was also influenced by the frequency of water exposure, and he found an incidence of 64% in individuals swimming three times a week.[33] Moore and colleagues observed that exostoses can develop in the presence of salt water as well as inland populations who swim in river water.[34]

It appears that the temperature of water and wind chill interaction is also important.[12] Specifically, Kennedy noted that external auditory canal exostoses were more common in communities lying between 30°8' and 45°8' latitude north and south of the equator, where the water temperature is below 19°C.[15]

More recently, Verhaegen has described that those who swim almost daily in water colder than 18–20° presented with a greater incidence of external auditory canal exostoses.[35] Kroon demonstrated in his surfer population that there was a higher prevalence of exostoses amongst the subjects who surf more frequently in the colder waters.[36] Interestingly, in Surfer's Paradise just south of Brisbane on the eastern Australian coast, ENT surgeons have noted that exostoses are more pronounced in the right ear compared with the left. It seems that the prevailing wind there is from the southeast and, as the surfers are paddling out from the shore to find a suitable wave, the right ear is more exposed to the prevailing wind and spray. The return journey to the beach, riding the surface of the wave, is much more rapid and the resulting exposure to the left ear is correspondingly reduced.

EXTERNAL EAR EXOSTOSES IN ARCHAEOLOGICAL POPULATIONS

Studies of exostoses in archaeological populations may reflect the littoral activities of early hominids as they utilised the marine environment in search of food or for other activities. *Homo* fossils have typically been found in the vicinity of rivers, estuaries and lakes. Barr attributed the high (18%) incidence of exostoses in the preserved crania of native mound- builders of central Tennessee to their frequent exposure to the river.[37] Verhaegen and Munro suggested that Pleistocene *Homo* populations similarly displayed littoral habits and that this is evidenced by the findings of pachyosteosclerosis.[38] In addition, CT evidence of ear exostoses in hominids during the middle Pleistocene era have been demonstrated in Spain by Perez and colleagues, confirming the hominids' aquatic lifestyle 1–2 million years ago (Figure 11.5).[39] Further studies have provided direct evidence of consumption of aquatic foods during the early Pleistocene era.[40,41]

External auditory canal exostoses have been found in several Neanderthals and some *Homo erectus* skulls, specifically *Homo erectus* skull X from Zhoukoudian, which suggests that they engaged in some type of aquatic activity.[15] Rightmire has also described a temporal bone with external auditory canal exostoses from Lake Ndutu in the Olduvai Gorge.[42] Verhaegen also presented convincing arguments that at least some of the Neanderthals had to be habitual divers. He described extensive and bilateral exostoses in the skulls of the middle-aged Shanidar I and Chapelle-aux-Saints males.[43] These fossil findings suggest that Neanderthals dived regularly and probably every day in cold rivers along which they lived.

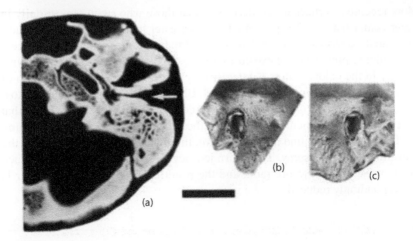

FIGURE 11.5   Temporomandibular lesions and ear exostoses of crania remains from the Sima de los Huesos Middle Pleistocene site, Sierre de Atapuerca, Spain.[39] (a) Transverse CT scan of cranium showing stenosis of left ear canal, (b) left ear canal, (c) right ear canal.

EAR EXOSTOSES AS A VITAL FOSSIL 'MISSING LINK'

Ear exostoses may appear a peculiar characteristic unique only to *Homo* and none of the other primates. It is suggested that a period of marine exploitation for hunting and other littoral habits could result eventually in significant adaptive changes to the anatomy and formation of exostoses.

Our understanding of the embryological development of the external ear canal demonstrates a logical explanation for the sites of development of exostoses and their pathogenesis. This ability to develop bony growths may appear redundant now, but it is highly likely that the presence of these growths constituted a selective survival advantage for early hominins during evolution.

Furthermore, a comparative study of adaptive changes in the external ear in other semi-aquatic mammals demonstrates parallel mechanisms for protection of the delicate hearing mechanism. It illustrates a logical explanation for their development in humans as a physiological modification rather than a pathological entity, providing an evolutionary survival advantage for early hominins in a marine environment.

Most of the other unique aquatic characteristics seen in humans but not in any other terrestrial mammal, such as hairlessness, subcutaneous fat, differences in the kidney and thermoregulation descent of the larynx and acquisition of voice, are all soft tissue adaptations and are therefore not preserved. On the other hand, the finding of exostoses and abundant bony remnants of aquatic foods provide vital fossil evidence of our aquatic past dating back to the Middle Pleistocene period 1–2 million years ago. Therefore, these findings can be considered crucial fossil evidence of humans' aquatic past and the validity of the waterside theory of human evolution.

Evolution of the Human Brain

12

> *No other land species has a brain capacity that
> remotely approaches the size and complexity
> of the human brain. The human brain is so
> incredibly superior to those of all other species
> that there ought to be a simple mechanism to
> explain the distinction.*
> Professor Michael Crawford (1992)

The evolutionary history of the human brain shows primarily a gradually larger brain relative to body size during the path from early primates to hominins and finally to *Homo sapiens*. Human brain size has been increasing, with a threefold increase from the early *Australopithecine* brains, about 3–4 million years ago which, at that time, were little larger than chimpanzee brains.[1] The increase in brain size started somewhat later, about 2 million years ago, from about 600 cm^3 in *Homo habilis* up to 1736 cm^3 in *Homo neanderthalensis* which is the hominin with the biggest brain. Since then, over the past 28,000 years following the demise of the Neanderthals, the average brain size has been shrinking: the male brain has decreased from 1,500 cm^3 to 1,350 cm^3 while the female brain has shrunk by the same relative proportion. However, the average body size of Neanderthals was larger than *Homo sapiens*, which corresponded to a larger brain volume (Figure 12.1).

Another essential element of brain evolution in humans is neurone (nerve) rearrangement.[2] Larger brains require more wiring, but more wiring can become more complicated. The brain has therefore become reorganized to become more efficient. The human brain contains about 100 billion neurons, which is equivalent to over 100,000 km of interconnecting nerves. Infolding of the human brain with extensive convolutions allows a great increase in internal volume and efficiency without expanding the overall brain size.

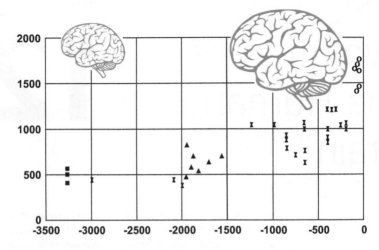

FIGURE 12.1   Increase in human brain size since 3,500,000 years ago. (Courtesy of the author.)

BRAIN STRUCTURE

From fossil records, scientists can infer that the first brain structure appeared in worms over 500 million years ago. The functions of the primitive hindbrain found in these fossils – what most neuroscientists call the proto-reptilian brain – included breathing, blood circulation and heartbeat regulation, balance, basic motor movements and foraging skills. According to a study done with mice, chickens, monkeys and apes,[3] the trend in brain evolution is that more evolved species tend to preserve the structures responsible for these basic functions.

A further study comparing the human brain to the primitive brain found that, in addition to the primitive hindbrain region, the modern human brain contains a new region of the brain developed about 250 million years after the appearance of the hindbrain. This region is known as the paleo-mammalian brain, and it deals with more complex functions including emotional, sexual and fighting behaviours. What this means is that evolution is the process of acquiring more and more sophisticated structures, not simply the addition of different structures over a long period of time.[4]

The two more recent major brain areas (the cerebrum and the cerebellum) are based on a cortical (two-layer) architecture. This means that the outer layer of the brain (grey matter) which is composed of layers (the number of which vary according to species and function) a few millimetres thick that are infolded into convolutions, in much the same way that a dinner napkin can be stuffed into a glass by folding it up. The degree of convolution is generally greater in more evolved species, which benefit from the increased surface area. This grey matter contains the nerve cell bodies, dendrites and nerve terminals where the nerves connect to each other (synapses). The deeper white matter contains the axons or nerve fibres which connect together with different parts of the brain and the spinal cord.

The cerebellum, or 'little brain', is at the back of the skull, behind the brainstem. Its purposes include the coordination of balance and fine sensorimotor tasks, and it may be involved in some cognitive functions, such as language. The human cerebellar cortex is finely convoluted, much more so than the cerebral cortex. The appearance of the interior nerve fibre tracts give them their name: *arbor vitae*, or Tree of Life (Figure 12.2).

The area of the brain with the greatest amount of recent evolutionary change is called the cerebrum, or neocortex ('new cortex'). According to research, the cerebrum first developed about 2 million years ago, which coincides exactly with the time when the human brain began to enlarge. It is responsible for higher cognitive functions – for example, language, thinking, and related forms of information processing. Most of its function is subconscious; that is, it is not available for inspection or intervention by the conscious mind.

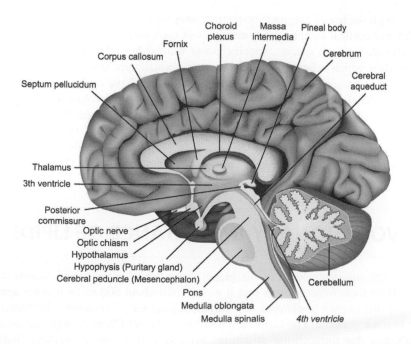

FIGURE 12.2   The cerebellum in humans. (Courtesy of medicalstocks/Shutterstock.)

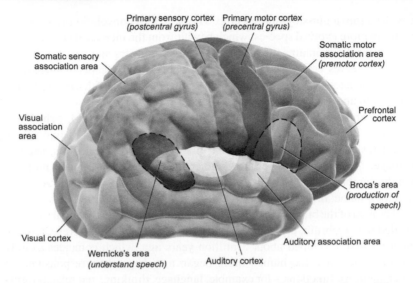

FIGURE 12.3   Areas of the human brain. (Courtesy of medical gallery of Blausen Medical, 2014.)

With the use of magnetic resonance imaging (MRI) and tissue sampling, different cortical samples from members of each hominin species have been analyzed.[5] In each species, specific areas were either relatively enlarged or shrunken, which indicate individual neural organizations. Different sites in the cortical areas can show specific adaptations. These have evolved depending on functional specializations of the different species and evolutionary events which result in changes in how the hominin brain is organized. For instance, during evolution, the enlarging frontal lobe, a part of the brain which is generally devoted to behaviour and social interaction, helps to predict the differences in behaviour between early hominins and modern humans (Figure 12.3).

EVOLUTIONARY IMPORTANCE OF LIPIDS

Although life originated on this planet about 3 billion years ago, scientists can infer from fossil records that it wasn't until about 600 million years ago (mya) that oxygen levels rose to a sufficient level for air-breathing life forms to become possible. The Cambrian fossil record from China around 580 mya indicates that intracellular structures evolved at this time containing lipid membranes.[6] With an oxygen-rich atmosphere, an entirely new biological process incorporating lipids and proteins to synthesize membranes became possible. Lipids and oxygen were therefore key determinants in this Cambrian evolution. Docosahexaeonic acid (DHA) is one of the most important of these marine lipoproteins and is essential for brain development. It provided the mechanism for the emergence of one of the key features of the first animals: new photoreceptors that responded to light and converted photons into electricity, laying the foundation for the evolution of other signaling systems, the nervous system and the brain.[4]

There is also clear evidence from molecular biology and fossil records that hominins took advantage of the marine food chain and that the omega-3 fatty acid DHA was important in determining neuronal migration, neurogenesis (new nerve formation) and the expression of several genes involved in brain growth and function. The same process was also essential for ultimate cerebral expansion in human evolution.[6]

DHA has been consistently preserved as essential in the nerve signaling systems of so many species including amphibians, fish, reptiles, birds, cephalopods, mammals, primates and humans despite wide genomic changes over 500 million years. This preservation confirms its vital importance in brain function. Incorporation of DHA into the developing brain from marine food has been shown to be greater than 10 times more efficient than its synthesis from the omega-3 fatty acids of land plant origin.

Another important aspect of DHA synthesis is that it depends on the size of the animal and the velocity of growth.[7] Small mammals, such as rats and squirrels, with their rapid metabolism, can readily convert the land-based polyunsaturated fatty acid precursor alpha-linoleic acid (ALA), found in plant leaves, grasses, walnuts and the meat of herbivores. Thus, the size of the brain, in relation to the body size of terrestrial mammals, decreases logarithmically with increased growth and size. For example, the rhinoceros, which weighs about one ton, only has a 350 g brain. Williams and Crawford have shown in their study of the Cape buffalo and the dolphin[8] just how difficult it is for a land-based mammal to develop a big brain because the pregnant mother would have little DHA to feed the fetal brain. A dolphin, with a 1.8 kg brain, is the closest to *Homo sapiens* in brain-to-body weight ratio and is exceptionally intelligent. Terrestrial mammals of a similar size, such as the zebra, however, have a brain a little larger than 300 g.

If we look at the comparative size of the brain in terrestrial mammals and aquatic mammals, there is a significant difference in mammals living in the two habitats (Figure 12.4). From the elephant down to the chimpanzee and whale down to the seal, the difference in brain size is considerable. But where do humans fit in? One can clearly see that they do not fit into the terrestrial mammal scale and even exceed the equivalent-sized brain of the similar-sized porpoise. This difference clearly shows that there is something in the aquatic environment which has a much greater influence on brain size and

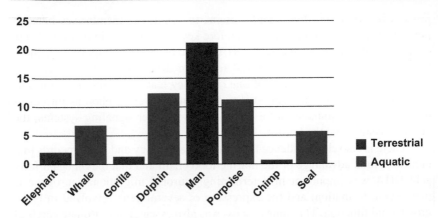

FIGURE 12.4 Comparative brain size of aquatic and terrestrial mammals. (Courtesy of the author.)

development than anything in the terrestrial habitat. Scientific evidence has shown that this is due to the dietary intake of DHA and arachidonic acid.[6-9]

DHA is the only omega-3 fatty acid in the brain, so it is likely that this is essential for gene expression and would have provided a selective advantage to a primate which separated from the chimpanzee in the forests and woodlands and sought a different ecological niche, benefitting from the marine and aquatic habitat. In this habitat, our ancestral apes would have had access to food rich in DHA as well as other essential nutrients for brain development such as iodine, zinc, copper, magnesium and selenium.[6]

Although the genome of *Homo sapiens* is 96% identical to that of the chimpanzee, our behavior is 96% different, and this difference is entirely due to our larger and more sophisticated brain. This cerebral expansion in hominins could not be explained by an evolutionary origin as savannah apes because the brain-specific nutrients were in poor supply inland but were abundant in a marine or aquatic diet.

Therefore, it is very unlikely that the evolution of a large complex brain in *Homo sapiens* could ever have happened on the savannah, and it is far more likely that this occurred in locations such as shores around salt water lakes of the newly formed Great Rift Valley or the sea coasts where there was abundant, DHA-rich, aquatic food available.[9] Iodine and other trace elements are also essential for brain development, and the marine food chain provided a good source. In contrast, in the Sudan and in other inland and mountainous regions, there are high levels of iodine deficiency which today is the most common cause of mental retardation, affecting 1.6 to 2 billion people. In England, the condition known as 'Derbyshire neck' was very common in the Peak

District in the Midlands, where iodine deficiency caused gross enlargement of the thyroid gland (which also uses iodine to make thyroxine). Today, in most countries with iodine supplements in the water, these problems are avoided.

Fossil evidence is incontrovertible on the subject of the exploitation of marine food at the time of the emergence of anatomically and behaviourally modern humans in Africa.[10–12] Chris Stringer from the Natural History Museum in London also supports coastal origins and migratory routes for *Homo sapiens*[13] from Africa to Europe.

Explanations for the evolution of the hominin brain mainly focus on developments in the epigenetic factors, which Darwin considered so important, including social environment such as group size, coalition formation and parental care, as well as physical environmental factors (climate, diet and food availability).[14] Although it is difficult to explain the selection pressures which determine the evolution of the human brain, neuro-imaging techniques for studying the comparative growth of the human and chimpanzee brain[15] have shown dramatic differences. In the 4-month embryo, the human brain is already twice as large as the chimpanzee brain, but after 22 weeks there is a sharp divergence when the human brain growth continues to accelerate, but the chimpanzee brain growth slows down. In early infancy there is also a rapid increase in brain volume in humans compared with chimpanzees.

Recent scientific evidence has demonstrated that evolution of the human brain, compared with that of other primates, reveals dramatic differences, not only in brain capacity but also in neural re-arrangement. The increasing size of the hominin skull from fossil remains over the past 2–3 million years only gives a crude idea of brain evolution. The main development in governing the unique increase in intelligence in hominins has been the expansion and internal nerve re-organization of the neocortex into a more complex structure, which is not evident in fossils. The difference in brain capacity and intellectual sophistication between terrestrial and aquatic and semi-aquatic mammals is clearly demonstrated, which can only be explained by the availability of an aquatic and marine diet rich in essential lipoproteins.

Food for Thought and the Cognitive Revolution

13

> *The measure of intelligence is the*
> *ability to change.*
> Albert Einstein (1879–1955)

The threefold increase in the size of the human brain over the past 2 million years is undoubtedly associated with a vast increase in intelligence compared to other primates and terrestrial mammals. As Charles Darwin wrote in *The Descent of Man:* 'The difference in mind between Man and the higher animals, great as it is, is certainly one of degree and not of kind'.[1] His theory of evolution, based on competition, adaptation and natural selection, is driven by the one vital force of survival in a particular habitat. As early hominins evolved larger brains and more intelligence, they were able to rely less on brute strength in hunting because they could use their cunning and increasingly sophisticated tool-making to be more successful in getting food. Throughout evolution, scarce energy resources are channeled into organs in the body which will enhance their chances of survival. In humans, this was the brain rather than muscle.

The increase in the size of the large brain in humans came at a price. The human brain accounts for about 2% of the body weight, but it consumes 20–25% of the body's energy when at rest. This contrasts to about 8% in other apes and primates and 3–5% in other non-primate mammals. This means that it needs more oxygen and consequently more blood.

'Brain size has increased about 350% over human evolution, but we found that blood flow to the brain increased by an amazing 600%', wrote Roger Seymour from the University of Adelaide.[2] 'We believe this is possibly related to the brain's need to satisfy increasingly energetic connections between nerve cells that allowed the evolution of complex thinking and learning'. The part of the brain

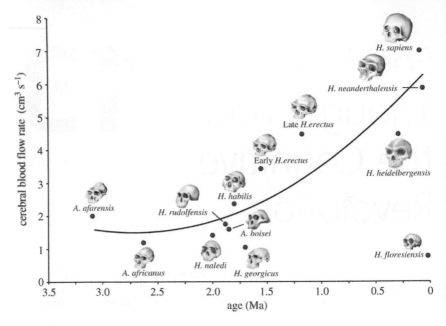

FIGURE 13.1 Cerebral blood flow in various hominin skulls. (Courtesy of The Royal Society.)

which has contributed to its increased size is the neocortex, which gets its blood circulation via the internal carotid arteries which enter the brain through the two carotid foramina in the base of the skull, one on each side. By measuring the diameter of the carotid foramen in hominin skulls, the team in Adelaide was able to estimate the cerebral blood flow at each stage in our evolution (Figure 13.1).[2]

From a medical point of view, it is of interest that the primitive hindbrain, responsible for control of the heart, circulation and respiration, which has remained largely unchanged in size throughout animal evolution, has a completely different blood supply compared to the large human neocortex. The blood supply to the hindbrain comes via the much smaller vertebral arteries which traverse up to the brain through small foramina in the cervical (neck) vertebrae (Figure 13.2).

We are familiar with seeing military guardsmen standing to attention for long periods at some official ceremony on a hot summer's day, and then fainting in the heat. As we shall see later in Chapter 17, these vertebral arteries are uniquely prone to temporary reduction in blood flow if the blood pressure falls for the simple reason that, unlike all other terrestrial mammals, we are bipedal. In an otherwise fit, young, 6-foot-plus soldier, extra effort is needed for the heart to pump blood up to the brain through these small vertebral arteries on a hot day when the blood circulation is slowed because of vasodilatation. This can

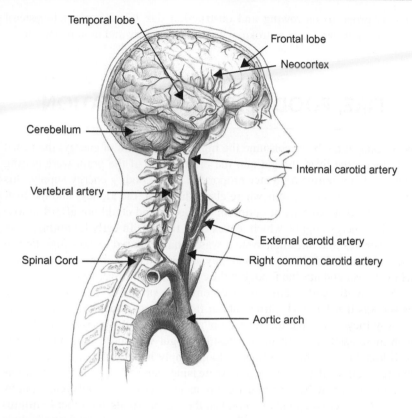

FIGURE 13.2   The internal carotid and vertebral arteries. (Courtesy of Fotosearch.)

result in temporary oxygen starvation of the vital hindbrain. The body's natural protective reflex is to faint so that the brain's circulation is restored again once the position of the head is brought down to the level of the heart.

Hypertension is another problem unique to humans because our mammalian circulation is not designed for a vertically aligned body. This would not happen in a primate or quadruped mammal where the heart and brain are more or less at the same level and the blood pressure to the brain is kept fairly constant. Patients on drugs to help control hypertension are more likely to suffer from momentary dizzy episodes when they are standing or walking because the medication is artificially keeping the blood pressure down.

These fainting or dizzy episodes are also common in bipedal humans who are prone to arthritic changes in the vertical spine where the vertebral artery canals can get narrowed. The carotid artery, which supplies the neocortex and higher functions, is not affected, however, by this vertebral artery circulation, but it is prone to narrowing and obstruction due to atheroma (cholesterol plaques) which can cause a stroke, with loss of sensory and motor functions.

FIRE, FOOD AND DOMESTICATION

Five organs in the body consume the highest proportion of energy: the heart, the kidneys, the liver, the gut and the brain. Thus, if the brain were getting bigger and consuming a greater proportion of the body's energy supply, this had to be diverted from somewhere else. The heart, the kidneys and the liver are continuously working at optimum efficiency and could not afford to lose their vital energy supply, which leaves only the gut. In early hominins, as in other primates, the intestinal system was initially geared to slow digestion of plant foods, as it did also later when they acquired a more omnivorous diet that included meat and marine food, but they still had slow digestion.

So how did early hominins overcome this problem? The only possible answer was that humans learned about fire and began to cook their food. In this way they were able to make their relatively tough vegetables and meat much more easily digestible; as a result the gut was shortened and required much less energy. This additional blood supply and source of energy could then be redirected to nourish the increasing demands of the brain. Some human species may have been using fire as long ago as 800,000 years, but by about 300,000 years ago *Homo erectus*, the Neanderthals and other hominins were using fire on a daily basis. They were able to widen their diet to include rice and potatoes which were much more edible when cooked.[3]

They also spent much less time eating food that had been cooked, unlike chimpanzees, which spend 5 hours a day chewing raw food.[4] Hominins were therefore able to devote more time to other activities such as hunting, socializing, making clothes and weapons, and other domestic duties, as well as artistic endeavours. Cooking also had the advantage of killing bacteria, parasites and other germs, which inevitably would have been beneficial to their health. Chewing and eating more digestible food also resulted in a change in dentition, jaw and facial structures associated with smaller teeth and less powerful masticatory muscles.

By the time the first *Homo sapiens* appeared around 200,000 years ago in East Africa, they were living a more domesticated life in small waterside communities where they evolved over a period of about 100,000 years. Fire was one of the most important factors that created a widening gulf between humans and other animals. Not only were they able to cook food, but fire also provided warmth in the winter months and at night so they didn't need their thick fur coat. In addition, fire afforded protection from predatory animals.

Most scientists agree that by 150,000 years ago, our ancestors living in East Africa were similar to us in appearance and were slowly beginning to become more independent and more able to adapt their environment and habitat to their own advantage rather than be dependent on it.

According to DNA analysis, there was a so-called genetic bottleneck around 70,000–75,000 years ago which reduced the human population to about 10,000 breeding pairs in equatorial Africa. This could have been due to disease or as a result of the Toba catastrophe, which was a massive volcanic eruption on the island of Sumatra in Indonesia that resulted in a huge ash cloud that blocked out the sun for a period of several years over a wide area of Asia and Africa. The eruption was 100 times greater than the largest volcanic eruption in recent history, the 1815 eruption of Mount Tambora, also in Indonesia, which caused the 1816 'year without a summer' in the northern hemisphere.[5] More recently, however, the significance of this eruption in affecting the hominin population in East Africa has been discounted.[6]

Whatever the cause of this genetic bottleneck, only the most resilient and resourceful animals, including humans, managed to survive, which must have proved a real evolutionary challenge for human survival. Perhaps this also was a powerful stimulus for more intellectual development, innovation and sophistication.

COASTAL MIGRATION AND WORLDWIDE DISPERSAL

Around 70,000 years ago, our ancestral humans in East Africa began the second exodus out of Africa by coastal routes to the Arabian Peninsula, before embarking on a journey northward and eastward to Europe and Asia, where they would live in the same ecological niche as the Neanderthals, *Homo erectus* and other humans. We know that when the African immigrants settled in Eurasia, they mingled and bred with other human populations and that modern-day humans are the progeny of these families. Europeans and people from the Middle East today have between 1–4% of Neanderthal DNA in their genome and, following successful fossil DNA analysis of a Denisovian finger, it established that up to 6% of modern Melanesians and Aboriginal Australians have Denisovian DNA.

Contrary to Victorian depictions of the 'barbaric and brutish' Neanderthals, these early humans were comparatively civilized. They had adapted to the harsh conditions of the ice age in Europe, used fire and tools, and were good hunters. From analysis of burial sites and fossil skeletons, some showing severe physical handicaps and old healed fractures, evidence shows that in their small communities they looked after their sick and wounded.

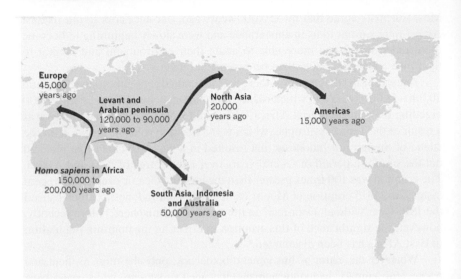

FIGURE 13.3   Worldwide dispersal of *Homo sapiens*. (Reprinted by permission from Springer *Nature*. "Climate and the peopling of the world." deMenocal et al. *Nature*, Volume 538, pages 49–50, 06 October 2016.)

By 45,000 to 50,000 years ago, humans had spread rapidly along coastal routes and had 'island-hopped' across Indonesia to Australia (Figure 13.3). Because of the frozen polar ice caps, sea levels were lower and, even though they had boats and rafts, sea crossings were shorter and less hazardous. Whatever the important factors that influenced the domination of *Homo sapiens* over their hominin cousins, *Homo neanderthelensis*, *Homo denisova* and *Homo soloensis*, whether it was physical or intellectual superiority, the last remains of *Homo soloensis* and *Homo denisova* date from about 50,000 years ago, and Neanderthals disappeared from France around 30,000 years ago and from southern Spain about 3,000 years later. Remains of the isolated dwarf-like *Homo floresiensis* in the Far East vanished around 12,000 years ago, leaving *Homo sapiens* as the sole surviving hominin species.[4]

THE COGNITIVE REVOLUTION

Although the brain size in hominins had progressively enlarged since the time of *Homo habilis*, genetic evidence has shown that it was only after the divergence of three genetic lines of *Homo sapiens* about 80,000–100,000 years ago that the intellectual advances really became apparent with increasing sophistication in tool-making and behaviour. The three main lines were distinguished by different mitochondrial haplogroups, L1 colonizing Southern Africa, L2 settling in Central and West Africa, and L3 remaining in East Africa.

The migration of *Homo sapiens* out of East Africa into Eurasia about 60,000 to 70,000 years ago seems to have been a stimulus for the rapid increase in intellectual development, with the appearance of figurative art, ritual burials, self-ornamentation with jewellery, musical instruments and increasing trade between different communities by 30,000–40,000 years ago.[7] This period saw the invention of seaworthy boats, oil lamps, bows and arrows, and needles.[4] This is associated with the Aurignacian culture typical of the Cro-Magnons, who seemed to have been more intellectually advanced than Neanderthals.

The half man/half beast lion-man found in Hohlenstein-Stadel Cave in Germany (Figure 13.4) dates from about 35,000 to 40,000 years ago and is the earliest known animal figurine ever found, carved from the ivory tusk of a woolly mamouth.[8] It does provide remarkable evidence of a new intellectual

FIGURE 13.4   The lion-man found at Hohlenstein-Stadel Cave, Germany, 35–40,000 million years before present (mybp). (Courtesy of World History Archive/Alamy.)

advancement of imagination, depicting a bipedal human-like lion that clearly does not exist in reality. This may have been representative of a developing religious culture, but certainly the early humans who drove the Neanderthals to extinction around 30,000 years ago were as creative and imaginative as modern humans. Other examples are paintings from the Chauvet Cave in France and the Venus figurines, and the earliest musical instrument, a bone pipe from Geissenklosterle, Germany, dated from around 36,000 years ago.[7]

A major aspect of this cognitive revolution was the unique evolution of human language and social communication. Animals of all descriptions have means of communication and co-operation, and some, like the bee and ant, have sophisticated ways of informing others about the whereabouts of food. Animals also have distinctive calls to warn others of danger. We are familiar with the melodic singing of blackbirds, but any sign of danger or disturbance and their song instantly changes to a shrill chirping call. Green monkeys will even distinguish between the imminent threat of a lion or an eagle with different calls.[7] A parrot can imitate almost any sound – from human speech to the ringing of a phone and a smoke alarm.

Marine mammals are particularly intelligent, and a recent study of 90 different species of cetacean has revealed that behaviour patterns of hunting techniques are passed down through successive generations. Their ability to communicate with whistles and squeaks through their sophisticated sonar mechanism and to develop tribal cultures, like ours, is evidence of their high intelligence. There are many examples of how both marine and terrestrial mammals work together to optimize resources and co-ordinate hunting skills.

For humans, the unique evolution of the human larynx and speech, together with intellectual development, was essential for a much more sophisticated level of communication which allowed us to communicate about how to find food, warn of danger, and enabled us to analyze data and use higher cognitive skills of imagination and reasoning. This led us to process information which could then be discussed with other members of the group to achieve an optimum plan.

The social brain hypothesis proposed by Robin Dunbar[9,10] suggests that human intelligence evolved as a means of surviving and reproducing in large and complex social groups. Our chimpanzee cousins usually live in troops of about 30–50 individuals; they form friendships, hunt and fight together and depend on each other for sharing food. Like other animals who live in groups or packs, the social structure is usually hierarchical, with a dominant individual, almost always a male. In contrast, it is likely that the evolution of language and verbal communication allowed early humans to form larger stable groups, which sociological research has shown to be a maximum of 150 people.

Scientists are not sure exactly what was the crucial factor in enabling early *Homo sapiens* to break away from this self-limiting ape-like existence, which meant that our ape cousins have remained unchanged socially and intellectually for over 20 million years, up to the present time. Humans, however, have totally changed. Most likely it was some form of microcephalin genetic mutation in the hominin brain as described by Lahn and colleagues.[11]

But what made this more likely to happen in our ancestral family rather than our ape cousins? It seems to me that the only logical answer was that, unlike any other primates or terrestrial mammals, we had much greater stores of the vital lipoproteins decosahexaeonic acid (DHA) and arachidonic acid, as well as iodine and other trace elements from our long-standing marine- and aquatic-based diet as a semi-aquatic mammal. The omega-3 fatty acid DHA is important in determining neuronal migration, neurogenesis (new nerve formation) and the expression of several genes involved in brain growth and function. It seems that a genetic mutation may have been the catalyst that switched our intelligence to be able to think and communicate in a much more complex way.

Although we lived on *terra firma* like our ancestral humans and other primates, what was vitally different for us was that we had evolved over 3–4 million years in a waterside, semi-aquatic habitat. Because of this habitat, some ancestral humans had the benefit of a predominantly omnivorous diet, including the precious elements of marine and aquatic food, which was to provide them with the nutritional neural requirements and enabled a chance genetic brain mutation that changed their destiny.

During this cognitive revolution, *Homo sapiens* began to explore communication with a wider social network outside their own group of 150 or so individuals for the purpose of trading and exchanging information about the outside world. There were so many things that they did not understand outside their own immediate ecological environment regarding the weather, the seasons, the sky and the galaxy that they began to believe in mythical and imaginary concepts of gods and deities from which religions have evolved. It is thought that the lion-man found in the Hohlenstein-Stadel Cave was a symbol of a lion-god mythology who would protect them. It is estimated that making this supernatural figurine would have taken more than 400 hours, which was a great deal of effort for a small community living in harsh conditions, for something that was not helpful for their physical survival. It suggests that it was something symbolic to help bonding in their community to overcome dangers and difficulties.[8]

GENETIC FACTORS CONTRIBUTING TO MODERN BRAIN EVOLUTION

Bruce Lahn, the senior author at the Howard Hughes Medical Center at the University of Chicago, and colleagues have suggested that there are specific genes that control the size of the human brain. These genes continue to play a role in brain evolution, implying that the brain is continuing to evolve. The genes were obtained from humans, macaques, rats and mice. Lahn and the other researchers noted points in the DNA sequences that caused protein alterations. These DNA changes were then scaled to the evolutionary time that it took for those changes to occur. The data showed that the genes in the human brain evolved much faster than those of the other species. Once this genomic evidence was acquired, Lahn and his team decided to find the specific gene or genes that allowed for or even controlled this rapid evolution.[11]

Two genes were found to control the size of the human brain as it develops: microcephalin and abnormal spindle-like microcephaly (ASPM). Under the pressures of selection, both of these genes showed significant DNA sequence changes. Lahn's earlier studies showed that microcephalin experienced rapid evolution along the primate lineage and eventually led to the emergence of *Homo sapiens*. After the emergence of humans, microcephalin seems to have shown a slower evolution rate. On the contrary, ASPM showed its most rapid evolution in the later years of human evolution once the divergence between chimpanzees and humans had already occurred.[11]

Each of the gene sequences went through specific changes that led to the evolution of humans from ancestral relatives. In order to determine these alterations, Lahn and his colleagues used DNA sequences from multiple primates, then compared and contrasted the sequences with those of humans. Following this step, the researchers statistically analyzed the key differences between the primate and human DNA to come to the conclusion that the differences were due to natural selection. The changes in DNA sequences of these genes accumulated to bring about a competitive advantage and higher fitness that humans possess in relation to other primates. This comparative advantage is coupled with a larger brain size, which ultimately allows the human mind to have a higher cognitive awareness.

Research suggests that the human brain is continuing to evolve and that variants of the two new genes that control brain development have swept through much of the human population over the last several thousand years. Although the size of the human brain has not changed appreciably over the past 200,000 years, even getting smaller over the last 28,000 years, it is difficult to evaluate how much these genetic adaptations have any effect on brain size or intelligence. Not everyone possesses these genes, which potentially inflames the controversial debate about whether this affects different groups of people.

Einstein remarked, 'The measure of intelligence is the ability to change' and since his lifetime in the early twentieth century, humans have experienced a dramatic exponential rise in change, particularly since the advent of the computer age, which I don't think even he could have envisaged. There is overwhelming evidence that the most significant factor in the physiological and anatomical changes in the hominin brain over the past 4–5 million years has been our unique high intake of a marine and aquatic diet rich in DHA and other vital elements. This distinguished our evolutionary adaptations from those of our primate cousins and subsequently enabled the evolution of human intelligence during the last 200,000 years.

Genetic variations and mutations have undoubtedly been influential in initiating changes which may have provided an evolutionary survival advantage, but it was our different waterside habitat and diet which gave our ancestors the essential nutrients for brain development, one of the epigenetic factors which Darwin considered so important.

We have seen how other unique physical and anatomical aquatic characteristics evolved in early hominins because of their different habitat, which provided a survival advantage over our other terrestrial cousins. Once *Homo sapiens* had become established as the dominant hominin throughout the animal kingdom, rivals fell away through natural selection. Domestication followed, then the agricultural revolution 10,000 years ago, the industrial revolution two centuries ago, and during the last 50 years the computer age. As Lahn has suggested, our intelligence is continuing to evolve, but what will be the price for humanity in the future? For those critics of the waterside theory of evolution who seem reluctant to accept an alternative scenario to the savannah theory, perhaps they should pause and reflect on Einstein's words.

The Human Larynx and Evolution of Voice

14

> *It seems that within the evolution of the human larynx may lie many of the secrets to understanding how we came to be.*
> Jeffrey T. Laitman and Joy S. Reidenberg

The most significant unique human characteristic which sets us apart from all other creatures in the animal kingdom is our ability to speak. It is also one of the most difficult aspects of hominin evolution to explain. Humans are able to convey very complex information, including facts and abstract ideas, using a highly variable rapid sequence of sounds. It does not require visual contact with the intended recipient(s) and can be carried out simultaneously with other tasks and activities. Humans are unique in being able to express this information in various ways, whether it is via sign or written language, but voice is the principal means of communication.

The intriguing question of how this unique characteristic of speech evolved in humans can only be answered by combining knowledge of our understanding of the anatomy and physiology of voice and sound production in humans and other animals, together with the archaeological record of fossils from early hominins. Only then can we hope to understand the evolutionary sequence of what happened and when. The question of why it happened, like other evolutionary changes, is simply answered by one word: *survival*.

A major limitation, however, in trying to understand the evolution of voice as well as other human characteristics unique to *Homo sapiens* is that, historically, scientists have assumed that our early hominin ancestors, like our ape cousins, were terrestrial mammals living on the savannah. In this land-based habitat, which is not dissimilar to that of our forest-dwelling relatives, it is difficult to determine what major factor(s) could have influenced the dramatic differences between us and our ape cousins. If, however, we accept the waterside ape theory and that early humans began to evolve as semi-aquatic rather than terrestrial mammals, we have a whole new dramatic dimension in evolutionary adaptations to aquatic and marine life which could have had a major impact on our ancestors' lives and influenced the evolution of these unique human features.

One crucial aspect of this complex topic is the importance of distinguishing between communication, human speech and human language, because they are all quite distinct.[2] A wide variety of animal species, including terrestrial and aquatic creatures and birds, have evolved various mechanisms to produce sound for the purpose of social communication,[3] as well as visual, sonar, smell (olfaction) and other sensory means.

We know that vocal communication in animals is produced though the throat and mouth, which evolved primarily for respiration and feeding. The vocal cords evolved in our distant marine and reptile ancestors many millions of years ago to prevent water entering the lungs. The permanent intersection at the back of the throat between the respiratory and digestive tract was a new feature evolved in mammals, but the descent of the larynx is unique to humans.

Many of the physical and anatomical features required for human speech were adapted specifically for this purpose, but the majority were *exaptations*, which is a term introduced by two well-known anthropologists, Stephen Jay Gould and Elizabeth Urba, and is used in evolutionary biology to describe a trait 'that has been co-opted for a use other than the one for which natural selection has built it'.[4] Evolution of the voice in adult humans is most likely an exaptation, which evolved in early hominin humans, and not in any other of our terrestrial primate cousins, as an adaptation to our aquatic waterside habitat to facilitate breath-holding during swimming and diving. After an evolutionary period of several million years, the anatomically new descended position of the larynx and the elongated mobile vertical tongue base, enabled by accompanying, genetically driven, neurological and intellectual changes in the cerebral cortex, may have led to a more versatile development of vocal abilities and the evolution of speech.

In humans and other mammals, sound is produced by air passing through the vocal cords, which are situated in the larynx. This may be either in inspiration or, more usually, during expiration of air from the lungs. The vocal cords are normally held open in a V-shape (the apex of the V being anterior) during respiration, but for vocalisation, the vocal cords are brought together and tightened so that air is forced through the narrow slit-like gap, causing the delicate membrane along the edges to vibrate and thus releasing a wave-like column of air which produces a sound, similar in fashion to the vibrating reed in a wind instrument. The pitch or frequency of the sound, known as the fundamental sound frequency (F0), depends on the tension and the length of the vocal cords. These sound waves ascend through the upper part of the larynx and vocal tract, where they are modified by movements and positioning of the tongue, lips, palate and teeth to form words and sounds. This is what we call voice.

THE UPPER AIRWAY AND DIGESTIVE TRACTS

Our primate cousins and many other animals use the larynx to produce sound,[5] but humans are unique because they have evolved a way of utilising the whole of the supraglottic ('above the vocal cords') vocal tract, in addition to the larynx, to produce speech. The supraglottic vocal tract includes the tongue, palate, teeth, and lips for articulation, as well as the nasal cavity and sinuses for resonation. The two most important aspects of this unique anatomical and dynamic difference in humans are the descent of the larynx and the multidimensional shape and mobility of the human tongue.

Although the nasal and oral airways are connected at the back of the mouth (oropharynx) in mammals and other terrestrial animals, some are considered obligatory nasal breathers, such as horses, rodents and rabbits. In these animals, the epiglottis rests above the soft palate, against which it forms an airtight seal (Figure 14.1). This allows the animal to breathe and eat at the same time without the danger of food or liquid entering the air passage and also maintains the sense of smell so that the scent of any predator in the vicinity can be detected.

Human infants are also considered obligatory nasal breathers and have a similar positional arrangement of the epiglottis and soft palate so that they can breastfeed and breathe at the same time. They are able to cry and produce sound, but they cannot produce voice until the larynx has descended, which is generally after six to twelve months of age. Most infants are able to breathe through their mouth if their nose is blocked. However, the rare condition of choanal atresia, where the back part of the nasal cavity is closed off at birth by a bony membrane, can prove fatal if it not recognised and surgically corrected. Infants may not be able to sustain mouth breathing for long

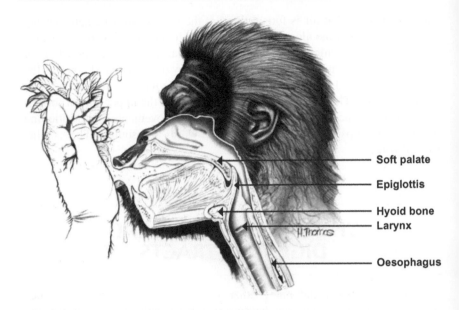

FIGURE 14.1   The horizontal tongue and close interlocking of the epiglottis and soft palate in the chimpanzee. (Courtesy of Jeffrey Laitman.)

periods of time due to weakness of the muscles required to elevate and seal off the nasal airway and open the oral airway.[6]

Older infants and children commonly suffer from enlarged adenoids, which can obstruct the posterior nasal airway, leading to Eustachean tube obstruction and middle ear problems. By the age of 2, however, the larynx has descended and the oral airway is well developed to avoid any airway problem. The unique process of laryngeal descent and separation of the palate and epiglottis in human infants, around the age of 6–12 months, where the vital airway is undergoing an important anatomical change of position, is thought to be associated with the danger and tragic occurrence of sudden infant death syndrome (SIDS).

Separation of the epiglottis and soft palate with descent of the larynx is not unique among humans and is seen in dogs, pigs, goats and deer, as well as in lions and other members of the cat family. In these nonhuman mammals, however, this is only a temporary lowering, for example, when they are emitting loud roaring, calls or signals. The hyoid bone stays in position and the tongue remains horizontal in the oral cavity, but it cannot act as a pharyngeal articulator as it does in humans (Figure 14.1).[7]

Philip Lieberman, whose central interest is in the evolution of human language, has stressed the importance of elongation of the tongue base associated

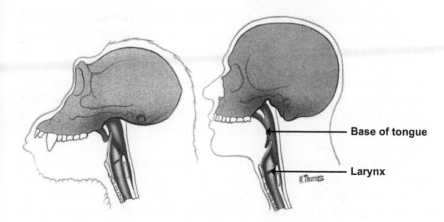

FIGURE 14.2 Chimpanzee (left) and the descended larynx in adult humans (right) with a greatly elongated base of the tongue. (Courtesy of Jeffrey Laitman. Modified from Laitman, J.T., *La recherche.*, 181, 17, 1165–1173, 1986 [Figs. 1 and 6].)

with permanent descent of the hyoid bone (Figure 14.2) as essential for voice production. This results in a curved tongue, with the horizontal part in the oral cavity and a similarly lengthened vertical part in the oropharynx, both of which are essential for human speech.[8] Scientific opinion is divided, however, about the reason for descent of the larynx in humans. It is not present at birth, and the advantage of having an interlocking soft palate and epiglottis permitting simultaneous breastfeeding and nasal breathing while the infant is totally dependent on his or her mother far outweighs any other potential evolutionary or survival benefit. It is an inherited feature, however, like male-pattern baldness, that develops later, in this case, once the baby has weaned off breastfeeding.

Once the infant has become more independent from its mother, the larynx begins to descend, disrupting the normal mammalian separation of the respiratory and digestive tracts during swallowing and thus increasing the risk of choking and aspiration of liquid and food. After puberty there is a second further descent of the larynx, but only in males. So what was the benefit of this dramatic change?

Some claim that it was for the purpose of enabling speech, but scientific evidence has convincingly shown that human speech evolved only during the past 50,000–100,000 years, and this unique anatomical reconfiguration in the vocal tract and upper aero-digestive tract with descent of the larynx must have evolved over several million years. John Ohala, emeritus professor of linguistics at Berkley University, argues that if descent of the larynx was an adaptation for speech, we would expect the male larynx to be far better adapted than adult females, but the reverse is true.[9]

Therefore, there must have been another basic evolutionary survival reason for descent of the larynx in humans over several million years and that the subsequent development of voice was a beneficial exaptation which evolved at a much later date in hominin evolution, during the cognitive revolution. The main anatomical change which seems to have taken place initially in the region of the back of the throat is elongation of the tongue base, which resulted in a doubling in length between the horizontal portion, which remained in the mouth, and the new vertical portion, with its attached hyoid bone and larynx below.

In anatomical and pathological terms, there are only three reasons why an organ in the body might elongate or move direction. The first is part of the normal growth process; for instance, the long bones in the limbs elongate because they have a growth centre at one end which produces new bone, causing the bone to grow longer. As a result, the attached muscles and other structures such as nerves and blood vessels also grow longer.

The other two reasons are because the organ is pushed from one end or it is pulled from the other end. Examples of the former are pulsion diverticula, such as the pharyngeal pouch or the diverticula in the condition known as diverticulosis. In these conditions, increased pressure in the lumen of the intestine causes the mucosal lining to be pushed out through the weakened muscle wall of the gut, causing elongated pouches outside the intestinal wall, at the back of the throat or in the colon.

An example of the latter is the traction bony protrusion of the mastoid process, a bony swelling one can feel just behind the ear. The powerful sterno-mastoid muscle on each side of the neck, which is the main muscle which turns the head from side to side, is attached to the base of the skull at the mastoid process superiorly and to the upper part of the collar bone and sternum (breast bone) inferiorly. Over a period of a few years the strong pull of the muscle causes gradual elongation of its superior bony attachment at the skull-base, forming the mastoid process.

The mammalian tongue lies horizontally in the floor of the mouth. Movements of the tongue are essential during the mastication process, when food is pushed around the mouth so that it can be chewed sufficiently before being swallowed. In considering the potentially bizarre reasons why the posterior part of the oral tongue might double its length in a completely different vertical direction from its horizontal position in the mouth, there are only two possible answers: either it was pushed down from above or it was pulled down from below. There is no logical reason why the hyoid and tongue base could be pushed down from above, which leaves the only possible solution: the larynx, hyoid and tongue base complex were pulled down inferiorly. But why?

For a savannah hunter-gatherer hominin evolving as a terrestrial mammal in a similar habitat to our ape cousins, there seems to be no logical explanation why one branch of the family and no other terrestrial mammal should gradually develop a two-dimensional, significantly elongated tongue with a descended larynx, without any obvious survival advantage, which is the basic reason for evolutionary change.

DIVING AND BREATH-HOLDING IN HOMINIDS

If we believe in the reality of the waterside ape evolving in a semi-aquatic habitat, it introduces a totally new dimension into the equation which seems to provide a simple and logical reason for descent of the laryngeal/tongue-base complex associated with swimming and diving in early hominins. We have seen how our ancestral hominid ancestors, as a result of the late Miocene drought and scarcity of food in their previous forest habitat, started foraging in the rivers, lakes, and estuaries and at the seashore where food was abundantly available. As they ventured into deeper water, developing bipedalism and other aquatic features, they learned to swim and dive, hunting for new sources of food.

This introduces the concept of breath-holding for deeper and more prolonged diving, which imposes new anatomical and physiological forces on the body, particularly in the circulation, the heart and the lungs, something we can readily see and monitor in modern free divers. Professor Erika Schagatay, who is an expert free diver, and her colleague, Andreas Fahlman, carried out a study to evaluate the diving performances of various aquatic, semiaquatic and terrestrial mammalian species.[10] The deep diver group such as the whales, dolphins and other cetaceans form a distinctly separate group, but several species can perform dives of intermediate duration and depth, and these make up the moderate diver group. A great number of air-breathing species, including humans, have more modest diving skills and form the shallow diver group.

While foraging in shallow water, humans may repeatedly dive to 20 minutes and may spend up to 60% of the time submerged in shallow diving, staying under water for up to 5 minutes at a time on one breath. The current maximum breath-holding time is amazingly 24 minutes 3 seconds and the maximum dive depth without fins is 101 metres. Trained and experienced swimmers can reach depths of up to 100 metres on a single dive. From this perspective, human diving capacity is well within the typical range of a semiaquatic mammal.

Swimming and diving data on free-living apes are not currently available, but a recent study on a pet chimpanzee[10] showed that it could breath-hold for an average period of 7 seconds while seated submerged in a pool. The chimp was observed to swim on the surface for up to 3 metres. Similar results were also recorded for an orangutan. It was thought unlikely that these mammals could voluntarily breath-hold for longer periods in the presence of rising P_{CO_2} levels. The voluntary diving and breath-holding abilities of humans would therefore seem to be unique hominin features amongst the primate order.

Human breath-hold divers are found around most non-arctic coastal areas in the world and include the Ama in Japan and the Bajau population in Indonesia, some of whom spend up to 9 hours diving on a daily basis.[11] In these communities the male divers were usually spear fishing and they were able to catch between 1 and 8 kg of fish, eels and octopuses, diving 3–9 hours per day. Women divers mainly collected clams, crustaceans and sea cucumbers. Children learn to swim and dive at an early age and learn to breath-hold from about six months old, often before they learn to walk.

In a typical spear fisherman's dive down to the seabed at 20 m, the heart-beat slows down to about 30 beats per minute and the lungs are compressed down to about one third of the normal volume, exerting a tremendous negative pressure on the trachea. Because the vocal cords are tightly closed to prevent water entering the lungs, the larynx (and with it the hyoid and tongue base) is forcefully pulled downwards towards the chest. The hyoid bone is firmly attached to the upper margin of the laryngeal (thyroid) cartilage by the thyro-hyoid membrane and therefore, with regular diving and hunting underwater, it seems to me that, over a period of time, the repeated downward traction of the larynx would gradually lead to vertical elongation of the tongue base muscle and hyoid (Figure 14.2).

The horizontal part of the tongue in the mouth would need to be strictly retained in its position because this was essential for eating. In due course, evolutionary pressures would probably favour divers who had adapted better to these pressure changes during diving and had a lower-positioned larynx. This eventually would become an inherited genetic trait because it presumably resulted in more efficient diving and hunting skills in the water and improved survival.

This gradual change in the anatomical shape and position of the tongue base and larynx probably evolved over tens or hundreds of thousands of years as a result of exploiting the marine and aquatic sources of food, but the change most certainly would have been firmly established well before the second exodus of *Homo sapiens* around 100,000 years ago. The stage had been set, however, in readiness for the eventual neurological and intellectual advances that were to come during the cognitive revolution, which permitted the evolution of human speech and language.

DELAYED DESCENT OF THE HUMAN LARYNX

In the human infant, like in all other terrestrial mammals, the larynx is in close contact with the posterior end of the soft palate. The nasal airway for breathing and the mouth and food passage are therefore separated to allow breastfeeding and breathing at the same time. At this stage in life, the baby is totally dependent on its mother for nourishment, care – for survival – occasionally crying when it needs attention or feeding.

It is only when the infant is sufficiently robust that it starts to become more independent from its mother and no longer needs breastfeeding. Weaning is the time when the infant has to start feeding autonomously, and from an instinctive evolutionary point of view, this is the stage at which the larynx starts to descend. This seems to be a genetically driven process; the survival benefit in early hominins, once they were eating independently, favoured those with a descended larynx. Over time, this unique anatomical human feature became an inherited characteristic. For an infant, optimal survival depended on the separated nasal and food passages for breastfeeding, but once the infant ate independently, the descended larynx, initiated by the evolutionary need to swim and dive for food, favoured survival thereafter.

EVOLUTIONARY CHANGES IN THE BRAIN FOR SPEECH AND LANGUAGE

Apart from the descended larynx and elongated tongue base, nonhuman primates have similar anatomical structures necessary for eating and breathing, but they have different neural control.[5] In apes and most mammals, the functions of eating and vocalising are separately controlled, and they are unable to perform both activities simultaneously. *Homo sapiens* is unique in being able to adapt and control these mechanisms designed for breathing and swallowing for the purpose of articulating speech. In order to achieve voice production, the human brain has evolved a specialized part of the cerebral cortex, called Broca's area, which is located in the lower portion of the dominant (usually left) frontal lobe. It controls motor functions involved with speech production, not only muscles of the larynx, tongue and vocal tract but also muscles controlling respiration.

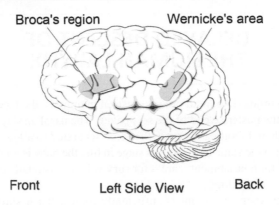

Broca's region   Wernicke's area

Front   Left Side View   Back

FIGURE 14.3 Speech and language areas of the brain. (Courtesy of the National Institute of Health.)

Language, on the other hand, is independent of speech, and the control of interpretation and expression of language has evolved in a different region of the brain, Wernicke's area, which is also in the dominant left hemisphere (Figure 14.3). It is quite distressing to see some patients with damage to Broca's area of the brain and with a normal-functioning Wernicke's area, which means that they can hear and understand language, but they cannot properly form words or speak fluently; they may become progressively mute and not be able to speak at all.

The important receptive part of Wernicke's area which processes language is called the *planum temporale*. It lies just behind the auditory (hearing) cortex. It was thought that the predominant asymmetry of the left side of this region of the brain was unique to humans,[12] but a study in chimpanzees by Patrick Gannon[13] showed that in 94% of animals (17 of 18), the left side of the *planum temporale* was significantly larger than the right. Thus, it is likely that the evolutionary origin of human language may well have already been lateralised to this part of the brain in our common ancestors over 7 million years ago. It would only be after the separation of our two lineages that further independent evolution resulted in the species-specific characteristics of human and chimpanzee communication and cognition.

One of the major differences in humans is that language is modality-independent, which means that the understanding or expression of language does not necessarily depend on the ability to speak or to hear. A congenitally deaf child can communicate perfectly well by learning sign language without actually having heard his or her own or another person's voice. Sign language provides another means of communication and is adaptable to any native language. Many species of animal use multimodal means of communication, with specific sounds and gestures to convey a meaning, but only humans, having lost one or two modalities of communication (e.g., hearing, sight), have the ability to switch instantly to another way of communicating (e.g., signing, typing, email).

With introduction of the concept of the aquatic/waterside ape, there does appear to be a logical explanation for the descent of the larynx in humans which later enabled the evolution of voice and language during the cognitive revolution 50–100,000 years ago. The current technological revolution has seen unprecedented rapid advances in instant communication and artificial intelligence, but sadly this may be associated with regression of some cognitive functions, and a deterioration in grammar and spelling and in the art of conversation and letter writing.

A recent study published earlier this year in the journal *Nature Communications* from University College, London,[14] showed that use of satnavs has resulted in switching off part of our brain. The hippocampus and prefrontal cortex areas of the brain are important for maintaining an awareness of our environment. The researchers found that the hippocampus, which is involved in both memory and spatial navigation, appears to encode two different maps of the environment: One tracks the distance to the final destination as the crow flies and is encoded by the frontal region of the hippocampus, the other tracks the 'true path' to the goal and is encoded by its rear region. During navigation tasks, the hippocampus acts like a flexible guidance system, flipping between these two maps according to changing demands. Activity in the hippocampal rear region acts like a homing signal, increasing as the goal gets closer.

In this study, volunteers who used maps and memory to get to their destination showed intense neural activity in these brain areas, in contrast to those who used satnavs, where there was no activity recorded at all. It seems that negotiating the complex narrow and unsystematic streets of London was much more taxing for the hippocampus compared with Manhattan, where the roads are designed in a systematic grid and one only has to make a decision to go right, left or straight on. Entering a junction such as Seven Dials in London, where seven streets meet, would enhance activity in the hippocampus, whereas a dead-end would drive down its activity. In the long term, one worries how much these important aspects of human communication will further decline and how much of our brain and intelligence will degenerate and become obsolete with increasing use of artificial intelligence (AI), robots and other modern technologies. There is no doubt that the potential benefits of AI are enormous, paving the way to solve highly complex medical challenges, advance scientific research, and better predict events and human behavior. However, as we become more and more reliant on AI to solve our day-to-day problems, is there a danger that this will alter the natural evolution of the cerebral cortex and human intelligence?

The ability of humans to communicate by speaking is unique in the animal kingdom. It is generally accepted that the descended position of the larynx in hominins is the one crucial difference in the vocal tract which allows continuous speech rather than momentary cries or calls. It seems impossible to explain why this change occurred in just one branch of the primate family who was evolving on the savannah rather than in the neighboring forest.

It is also evident that this change in position of the larynx gradually evolved over a period of several hundred thousand years and must have occurred for one reason only – to improve that family's ability to survive in a different habitat. It did not come about for the purpose of evolving speech, but as an earlier exaptation, because speech evolved much later over a relatively short period of time, during the cognitive revolution.

Several historical theories about evolution of voice have been proposed, based on fossil evidence and the assumption that early hominids evolved on the savannah. However, they do not seem to be credible because they do not conform readily with established physiological or anatomical principles. I believe that the only logical physiological explanation for descent of the larynx was because of repeated negative traction forces pulling the larynx and tongue base downwards during diving, to enhance the underwater hunting skills of early hominins.

This anatomical adaptation, in the same way as ear canal exostoses, expansion of the paranasal sinuses and elongation of the nasal skeleton, were all evolved over several hundred thousand years for the one purpose of improving the swimming and diving skills of early hominins and, ultimately, their survival. Much later, the lower position of the larynx and changes in the vocal tract subsequently allowed greater vocalization skills and the acquisition of speech.

Obstetric and Neonatal Considerations

15

The advent of the modern birth canal,
the shape and alignment of which require
foetal rotation during birth, allowed the
earliest members of our species to deal
obstetrically with increases in encephalization
while maintaining a narrow body to meet
thermoregulatory demands and enhance
locomotor performance.
Laura Tobias Gruss and Daniel Schmitt (2015)

So many unique differences between human babies and the newborns of other primates cannot simply be explained by the fact that our branch of the ape family evolved as terrestrial mammals, similar to other primates, but on the savannah rather than in the forests. There does not appear to be very much difference between these two adjacent land-based habitats, with similar access to food sources following the late Miocene drought, similar predators and other local influences for us on the savannah and apes in the forest.

There must have been other, very significant evolutionary epigenetic forces, which Darwin considered so important, over time that resulted in such major differences in length of gestation, degree of maturity at birth, intrauterine brain development, lung and heart maturation, lacuna, vernix caseosa and the length of postnatal maternal care. These unique human modifications are not seen in any other primate or terrestrial mammal, but they are seen in various forms in aquatic and semi-aquatic mammals. Their presence in human babies can only be explained logically if we accept the waterside ape theory and the substantial influence of an evolutionary period as a semiaquatic mammal.

THE OBSTETRIC DILEMMA

Being born as a human is no easy matter – both for the newborn and for the mother! The main problem relates to the size of the baby's skull which makes her or his passage through the birth canal fairly hazardous and all too often life-threatening. For chimpanzees and other primates, giving birth is much easier as their brain is much smaller. The newborn chimpanzee has a brain size of around 155 cc which is about half the size of a human full-term baby, although the birth canal is roughly of similar dimensions to that of humans. The shapes of the human and chimpanzee pelvis, however, are quite different because of the evolution of bipedalism, as evident in Lucy from 3.4 million years before present (mybp) and other *Australopithecines*.

We have seen how foraging, swimming and diving for marine and aquatic food resulted in a gradual change in the orientation of the lower limbs, from a quadruped to a more streamlined bipedal gait, in our earliest human *Australopithecine* ancestors. The need to keep the head and airway above water level and buoyancy of the water allowed progressive rotation of the pelvis and spine to a more vertical position. In other quadruped primates, the pelvic cavity is deeper, but the change to a vertical posture in hominins meant that a shorter and wider supporting pelvis was required. However, the demands of walking and mobility on two legs instead of four added constraints on how wide the pelvis could be. In addition, the necessity to bear the weight of the whole body on two legs rather than divided among four, one at each corner, meant that the musculature of the lower limbs and their attachments to the pelvic bones needed to be much stronger.

The resulting obstetric dilemma in early *Australopithecines*, balancing the need for a wider pelvis for childbirth and a narrower and stronger pelvis for bipedal mobility, inevitably led to evolutionary modifications and compromises for ultimate better survival, easier and safer childbirth and successful procreation. Apart from modifications in the *Australopithecine* pelvis, there had to be changes in the fetal skull to accommodate the enlarging brain and also in shortening the length of gestation so the baby could be born before the brain became too large.

FONTANELLES AND SKULL SUTURES

The vault of the primate skull is composed of four paired bones: the frontal bones between the forehead and the top of the skull, the parietal bones behind these, the occipital bones at the back of the head and the temporal bones at the

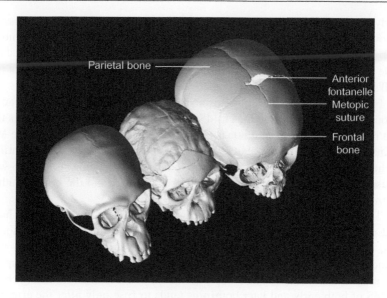

FIGURE 15.1 Three-dimensional computerized tomography (CT) reconstructions of a human infant (right), chimpanzee (left) and a 3-million-year-old *Australopithecus africanus* skull of a 4 year old. (Courtesy of M. Ponce de León and Ch. Zollikofer, University of Zurich.)

side, above the ears (Figure 15.1). They are joined by sutures which are initially fibrous during the growth period, but they eventually undergo bony fusion. In addition, at the junction of the frontal and parietal bones above the forehead and further back, at the junction of the parietal and occipital bones, are the so-called anterior and posterior fontanelles. The anterior fontanelle is sometimes referred to as the soft spot felt in the midline at the top of the baby's forehead.

In order to accommodate the rapid growth in the human fetal brain up to 400 cc at birth and the explosive doubling in size during the first two years of life, up to about 800 cc, the skull bones need to be able to enlarge and expand, and they can only do this because the sutures and fontanelles are not firmly fixed. This flexibility of the skull is also beneficial during childbirth, allowing the skull to 'mould' as it passes through the birth canal.

In contrast, the chimpanzee infant skull does not need to expand very much and the sutures and fontanelles become fused shortly after birth. A recent study has shown that, although the early hominin fetal skull was much smaller than ours today, the delay in fusion of the sutures and fontanelles was evident as early as 3 million years ago.[1] A team led by anthropologist Dean Falk looked at a key marker of skull fusion in a large number of fossil early humans, modern humans, chimpanzees, and bonobos. The study involved a new analysis of the nearly 3-million-year-old Taung child, a 4-year-old *Australopithecus africanus* discovered in South Africa in 1924 by the legendary anthropologist Raymond Dart. The fossils include a face, lower jaw, and a natural 'endocast' of the skull made up of the rocky material that filled it. The endocast preserved many of the skull's features, including the sutures between its bones.

Falk and his team used computerized tomography (CT) scans of the Taung endocast to closely examine the child's metopic suture (MS), which forms the joint between the two frontal bones (Figure 15.1). In human children, the MS begins to fuse on the end nearest to the nose, closing like a zipper until it reaches the anterior fontanelle. The team found that the Taung child's MS was unfused, even though its brain was only about 400 cc and the brain of adult *Australopithecus africanus* is only about 460 cc.

The researchers compared the Taung child's MS to that of several hundred chimps and bonobos, more than 1000 modern humans, and 62 hominins, or ancient humans, including *Australopithecines*, *Homo erectus*, and Neanderthals. A clear pattern emerged and showed that the MS of chimpanzees and bonobos fuses very shortly after birth, whereas, like the Taung child, the MS of both early and later hominins tends to fuse only after the eruption of the first molars, at 2 years of age or later.

Falk and her colleagues suggested that the obstetric dilemma may have already been a problem for early hominins as they began to walk upright. Although the brain of an adult *Australopithecus* is small compared with modern humans, it was already about 22% larger than that of a chimpanzee. The early hominin brain may also have begun to acquire the accelerated growth seen in the postnatal period, particularly in the frontal lobes where the neocortex was beginning to evolve into a more convoluted structure.

BRAIN SIZE AND PERINATAL CONSIDERATIONS

One crucial factor, therefore, in the determination of human brain size has been the limitation in the size of the female hominin pelvis. By 2.4 million years ago, *Homo habilis* had appeared in East Africa, the first known human species and the first to make stone tools. Our knowledge of the complexity of behaviour of *Homo habilis* is not limited to stone culture, as they also had habitual therapeutic use of toothpicks.

The use of tools conferred a crucial evolutionary advantage, and required a larger and more sophisticated brain to co-ordinate the fine hand movements required for this task. But the evolution of a larger brain created a problem for early hominin females because a larger skull requires a larger birth canal. However, if the birth canal became too wide, her pelvis would be broader and she would lose the ability to run. In our earliest upright ancestors, the *Australopithecines*, there had to be fundamental alterations in the pelvis compared with our quadruped primate cousins to facilitate bipedal locomotion. They had a wide platypelloid birth canal with flared iliac bones, which was maintained for 3–4 million years with little change until the emergence of *Homo sapiens* in Africa and the Middle East about 200,000 years ago, when the narrow anatomically modern pelvis with a more circular birth canal evolved.[2]

This major change appears to reflect selective pressures for increased neonatal brain size and for a narrow body shape associated with heat dissipation in warm environments. The advent of the modern birth canal, which requires fetal rotation during birth, allowed the earliest of our species to optimise the dual requirements of increased brain size whilst maintaining a narrow body to meet thermoregulatory demands and locomotor performance.[2]

The solution was to give birth at an earlier stage in fetal development, before the skull grew too large to pass through the birth canal. However, this imposed another problem of needing to care for an immature infant. This meant that the mother needed to spend a prolonged period of time confined in one place and relatively immobile, bringing up, feeding and protecting her child from predators and teaching it survival skills whilst waiting for the child to walk and become more independent. Meanwhile, the males would be hunting on land and in the water for food. And so, humans became more confined to a campsite waterside habitat, dependent on tool making and on agility for hunting on land and in the water. They became more intellectual and sophisticated, developing domestic activities and becoming less reliant on brute strength and size.

LANUGO AND FAT BABIES

Several distinctive features about human babies are not seen in other terrestrial mammals, and many scientists feel that these may be associated with adaptations to our semi-aquatic origin. During the third month of gestation, the human foetus begins to grow a fine coat of hair known as lanugo, which covers the entire body by the fifth month, only to disappear almost completely by 36 weeks.

During this mid-trimester period, once all of the major organs have developed, most of the baby's energy is diverted into the production and laying down of sub-dermal fat, which is uncharacteristic of terrestrial mammals but is typically seen in aquatic and some semi-aquatic mammals.

During the third trimester, the fat stores increase dramatically, from 30 grams to over 400 grams; at term, the average amount of fat accounts for about 16% of body weight, compared with 3% in newborn baboons. The seal is the only other documented mammal which has an equivalent high proportion of fat at birth.

This change in the composition of the skin during the gestation period, with loss of the coat of lanugo hair and replacement by a thick layer of subcutaneous fat, is thought to reflect the genetic sequence in our evolutionary history. The initial development of a coat of hair is similar to that in other terrestrial mammals to protect and insulate them on land. But then to lose that and immediately replace it with a much more efficient water-insulating and buoyant subcutaneous fat layer in preparation for birth would suggest that there was something very different about our evolutionary history compared to other primates and terrestrial mammals. The major difference was that the fetal human skin was switching *in utero* from a traditional terrestrial covering of primitive hair to the unique requirements that would be needed as a semi-aquatic buoyant and insulated baby.

VERNIX CASEOSA

One of the other curious features about human babies when they are born is the presence of a white cheese-like substance, called vernix caseosa, coating the baby's skin. In the womb, the baby is bathed in amniotic fluid for 40 weeks and if it didn't have a protective waterproof coating of vernix, the skin would become wrinkled and inflamed. The amount of vernix decreases as the baby's waterproof skin matures in the womb, but in premature infants the vernix may be quite thick (Figure 15.2). The vernix helps to protect the baby in the womb from infection and also contributes to softness of the baby's skin. It may also help to lubricate the skin surface during delivery. The reason that full-term babies have little or no vernix is that they have ingested it with amniotic fluid during the gestation period, as it also has a protective function in the intestine.

The only other mammal known to be born with a covering of vernix caseosa is the seal, in which there does seem to be a correlation between the thickness of the vernix and the time elapsed before the newborn seal, which is born on land, enters the water. The hooded seal baby pup does not enter the water for over a day and its vernix coat is very thin, the grey seal is in the water after a few hours and its covering is thicker, but the harbour seal goes into the water within 30 minutes and its coat is the thickest, although it is still thinner than in humans.

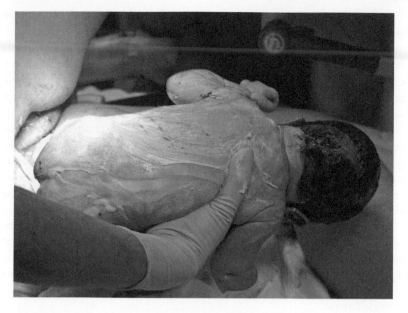

FIGURE 15.2   Vernix caseosa in newborn infant. (Courtesy of Llapissera.)

AQUATIC ADAPTATIONS OF HUMAN BABIES

The idea of a human 'water baby' was first introduced in the 1930s by Myrtle McGraw, and her work quickly inspired a widespread interest in swimming in newborn babies and infants. One can easily observe that even in the first few months of life, human babies are able to float, swim and dive quite safely. Over the past 40 years there has been an increase in the practice of water-birth and baby-swimming courses. None of this type of behaviour has ever been seen in other primate babies such as the apes, chimpanzees, gorillas or bonobos.

Animals naturally choose to give birth in the safest and easiest environment, and among mammals only the fully aquatic whales, dolphins and other cetaceans and the semi-aquatic mammals such as the hippopotamus and sea otter give birth in the water. However, until recently, human natives in coastal areas would naturally choose to give birth in the sea, but the practice stopped when missionaries thought it was wrong and made them stop.[3]

Dental fossil records for early humans and hominins, including *Australopithecines* and members of *Homo*, reveal that these species have a quiescent period of growth. This is a period in which there are no dental eruptions of adult teeth. At this time the child becomes more accustomed to social structure, and development of culture. It has an extra advantage over other hominins, devoting several years to developing speech and learning to co-operate and integrate within a community.

This period is also discussed in relation to encephalisation. It was discovered that chimpanzees do not have this neutral dental period, and scientists suggest that a quiescent period occurred in very early hominin evolution. From the models for neurological reorganisation, it can be suggested that the cause for this period, dubbed middle childhood, is most likely for enhanced foraging abilities in varying seasonal environments.

So many differences in pregnancy, pelvic anatomy, childbirth and neonatal development between humans and other ape primates are difficult to reconcile with a similar land-based savannah habitat for early hominins. There does not appear to be any logical evidence or any parallel evolutionary progression in other land species, and all of the available data are supportive of an aquatic/waterside habitat.

Marine Adaptations in the Human Kidney

16

*The multiple medullary pyramids of the
human kidneys probably evolved as an
adaptation to a coastal marine ecology.*
Marcel Williams (2011)

THE GREAT RIFT VALLEY

It is well established that all of the major events in early hominin evolution during the last 10 million years took place in East Africa in the region of the Great Rift Valley, which stretches from the Afar Peninsula, where the Red Sea joins the Indian Ocean down the eastern side of the continent, ranging from 40 to 60 km wide, as far as Mozambique in the south. During this time, there were dramatic changes in the climate and landscape in this part of Africa which undoubtedly had a major influence in the emergence of hominins. It changed from a relatively flat uniform region covered with subtropical forests to a mountainous area with forests as well as areas of desert.[1]

Tectonic activity in this region caused gradual separation of the Nubian and Somali East African plates and resulted in the formation of the Great Rift Valley. The Great Rift Valley is composed of two main rift valleys, the Western and the Eastern rift branches (Figure 16.1). From the north, inundation of the Eastern Rift from the Gulf of Aden led to the formation of an inland sea and relatively shallow and small salt water lakes. There were also many volcanoes associated with the Eastern Rift. Further inland, the Western Rift had

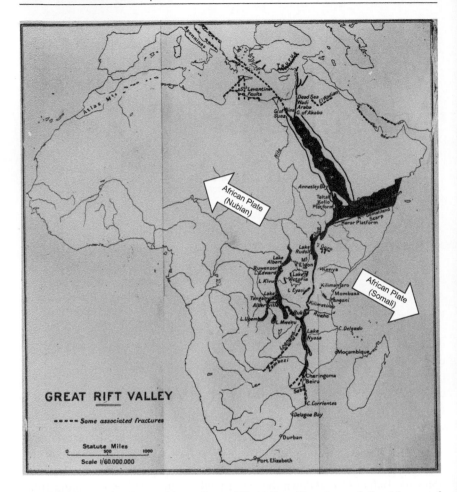

FIGURE 16.1 The East African Great Rift Valley. (Courtesy of the Library of Congress, Prints and Photographs Division. Reproduction number LC-DIG-matpc-00451; digital file from original photo.)

few volcanoes but much deeper and larger, mainly fresh water lakes. Evidence suggests that the lake basins of the Eastern and Western Rift branches were often transient and highly sensitive to local climatic changes and that these periodic shifts may have been influential in dictating events in hominin evolution, including hominins' eventual exodus out of Africa.[1]

The tectonic and geological history of this part of Africa has provided ideal conditions for the preservation and ultimate finding of vital fossils of early hominins and their tools, and these have allowed us to piece together the four main stages of human evolution.[2] The first of these was the appearance of the earliest hominins of the genera *Sahelanthropus*, *Orronin* and *Ardepithecus* between 4 and 7 million years before present (mybp); the second and more widely recognised stage was the period of the *Australopithecus* around 3.5–4 mybp and *Paranthopus* around 2–7 mybp. The earliest member of the *Homo* genus appeared about 1.8–2.5 milllion year ago, and finally anatomically modern humans (AMHs) appeared around 200,000 years ago.

Most of the early fossil specimens from before 5.5 million years ago are only fragments of skull or dental artefacts, and although the size of their skull was similar to that of the chimpanzee, the lack of post-cranial remains make it extremely difficult to reconstruct their lifestyle or to determine whether or not they were bipedal. More complete skeletal remains have been found from *Ardipithecus ramidus* around 4.4 mybp which suggest that, although the brain and body size were similar to modern chimpanzees, post-cranial evidence suggests a semi-arboreal and primitive bipedal mode of locomotion.[3] The fauna and vegetation associated with *A. ramidus* do suggest a woodland matrix with a significant waterside habitat.[3,4]

Findings of the more recent *Australopithecus afarensis* and almost complete skeletons of Lucy from 3.7 mybp have conclusively shown that early hominins were bipedal by that time, which marked the definite transition from the more primitive *Ardipithecus ramidus* 500,000 years earlier. One of the last *Australopithecines* identified is *A. anamensis* from 2.5-million-year-old deposits found in the Awash Valley, and it is characterised by a longer femur than other *Australopithecines*. This feature was interpreted by Green and colleagues[5] as suggesting a more efficient walking style with longer strides, but it could also be associated with stronger lower limbs to support a bipedal body and a more efficient swimming and diving technique.

The third and perhaps the most important stage of hominin evolution saw the appearance of *Homo habilis*, the 'handyman', around 1.8–1.9 mybp, who had a similar appearance to *Australopithecines* but used tools associated with a slightly larger brain. The first exodus out of Africa into Eurasia happened about this time, and shortly afterwards *Homo erectus* followed, with major changes in morphology and lifestyle, increased brain size and a shape becoming more similar to AMHs. The brain size of *Homo erectus* and *H. ergaster* were 80% larger than that of *Australopithecines* and 40% larger than that of *Homo habilis*. The final stages of human evolution were associated with the appearance of *Homo heidelbergensis* around 800,000 years ago and AMHs 200,000 years ago.[1]

HOMINID KIDNEY ADAPTATION TO A WATERSIDE AQUATIC HABITAT

The importance of salt in the cells and body fluids of all living creatures cannot be overemphasized. Life on earth has evolved originally from sea creatures, and the presence of salt (sodium chloride), which constitutes 90% of the entire ocean's mineral content,[1] is essential for the normal metabolic function of most cellular processes. Although they cover more than 70% of the world's surface, the oceans make up 99% of the earth's total living space because of their depths.[2] Our blood contains roughly the same percentage of mineral content as the ocean but is four to five times less salty than the ocean.[3]

Mammals have evolved on earth over a period of 50 million years and have adapted to life on land as terrestrial animals, but some have ventured back to the sea as aquatic mammals, such as the whales, dolphins, porpoises and other cetaceans. There are also a group of semi-aquatic mammals including seals, hippopotamuses, otters, and beavers who live partly on the land but spend much of their time hunting for food in the water.

It is acknowledged that hominin speciation (formation of new species) events and changes in brain size seem to be statistically linked to the occurrence of transient deep-water lakes in the Great Rift Valley. The emergence of *Homo erectus*, with their big brains, around 2 mybp was co-incidental with the period of maximal lake coverage.[1] Many of the lakes in the Eastern branch of the Rift Valley were alkaline from inundation of the sea from the Afar Peninsula, and several of the early hominin fossils, such as those from the Olduvai Gorge, were found in this region.

In mammals, the kidneys function as the principal organ for the excretion of ingested and metabolically produced water, salts and nitrogenous waste.[6] Internally, the outer part of the mammalian kidney is known as the cortex, which surrounds the inner medulla. Blood enters the kidney cortex where it is filtered by the glomeruli, allowing the blood cells and larger molecules of protein and fat to remain in the circulation whilst the water and smaller molecules of salt and nitrogenous waste filter through the glomerular membrane. This waste fluid then passes into the medullary part of the kidney where the water is reabsorbed back into the circulation, leaving the salt and waste products to pass out of the body in the urine.

The structure of the kidney in humans shows a multi-pyramidal morphology, which is almost totally unique among primates and terrestrial mammals

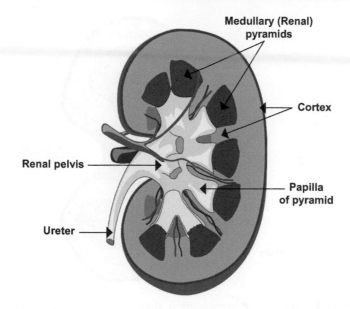

FIGURE 16.2   The human kidney with multiple pyramids. (Courtesy of the author.)

but is nearly universal in marine mammals (Figure 16.2). In salt water environments, the multi-pyramidal structure functions by increasing the amount of salt and nitrogenous waste excretion through increasing the surface area between the cortex and the medulla, in a similar way to the involuted structure of the cortex of the human brain.

In fresh water environments, multiple pyramids do not seem to serve any function and most of the terrestrial mammals have a uni-pyramidal morphology (Figure 16.3). Other terrestrial mammals with multiple pyramids such as elephants, bears, and rhinoceros have semi-aquatic ancestors which frequented marine habitats. Others, such as the Bactrian and Arabian camel, have multi-pyramidal kidneys that evolved because of a highly concentrated intake of salty water and plants (Figure 16.4).

Other aquatic adaptations in the human body, such as salt-excreting eccrine sweat glands and the excretion of salt tears, are similar to other marine and aquatic mammals. They are further evidence of an evolutionary adaptation to a waterside or coastal marine habitat rather than a relatively dry terrestrial environment.[6]

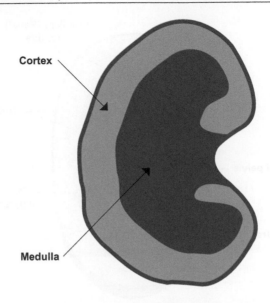

FIGURE 16.3 Longitudinal section of a sheep kidney, displaying the typical uni-pyramidal morphology found in most terrestrial mammals. (Courtesy of the author.)

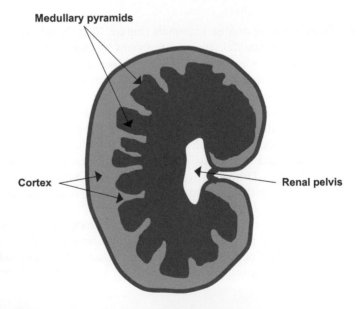

FIGURE 16.4 Longitudinal section of the kidney of a dromedary camel (*Camelus dromedarius*), displaying its renal pyramids. (Courtesy of the author.)

Scars of Evolution

17

We must, however, acknowledge, as it seems to me, that man with all his noble qualities still bears in his bodily frame, the indelible stamp of his lowly origin.
Charles Darwin (1809–1892)

MAMMALIAN EVOLUTION

If we go back a long way in earth's history, mammals evolved from a branch of the reptile family called the synapsids which have been around for over 200,000,000 years. Reptiles are cold-blooded animals and rely on external heat to raise their body temperature sufficiently for them to become active. At night and in cold weather, everything slows down while the body digests and processes food. Thus, they have to wait until the temperature warms up during the day before they can again be active, usually for short periods of time sufficient for them to find another meal.[1]

Evolution to a mammalian form required substantial changes because mammals are warm-blooded and generate heat internally by metabolising food as an 'internal combustion engine'. This keeps their body temperature fairly constant so they can remain active for much longer periods, although they too need some downtime to rest the muscles and divert energy to internal organs such as the brain, kidneys and the digestive system. In addition, mammals had to develop an efficient way of keeping the heat in, by having an insulating fur coat or layer of fat or blubber.

In order to achieve this increased activity, their metabolism needed to be more efficient, and they required more food as well as oxygen. Adaptations to strengthen the heart with four chambers and expansion of the lungs were needed to provide a more effective circulation, and changes were also required to excrete metabolic waste products more efficiently. This meant that the kidneys and the urinary system had to be separated from the digestive tract and faecal waste. Most of the early mammals were quite small, similar to members of the shrew family, some of whom, including the platypus, still retain the combined excretory tracts.[1]

Another major difference between reptiles and mammals is in their reproductive system. Reptiles lay eggs, and when they hatch the baby hatchlings are fairly mature and have to fend for themselves. Mammals, on the other hand, evolved a system whereby the young are protected inside the mother for a longer period of physiological development. After they are born, baby mammals also benefit from an extended period of parental care, taking nourishing milk from newly evolved maternal mammary glands. The additional protection and longer period of development meant that they did not have to survive and fend for themselves immediately after birth, but they needed time and resources to evolve into a more complex animal.

Whales, dolphins and other cetaceans took to the sea about 100 million years ago, where they were able to evolve a much more highly developed brain and nervous system because of the ready availability of the two vital lipoproteins essential for neural growth: docosahexaeonic acid (DHA) and arachidonic acid. Two to four million years ago, these lipoproteins became such a crucial element in the successful semi-aquatic evolution of early hominins.[2]

HOMININ EVOLUTION

Raymond Dart announced in January 1925 that he had found the skull of a 4-year-old 'Taung child' in South Africa whom he named *Australopithecus africanus* and claimed that he had found the 'missing link' between apes and humans. He was denounced and ridiculed for not recognising the skull of a young chimpanzee, but eventually in 1947, after Broom's finding of an adult *Australopithecine*, Dart's opinion and findings were gradually accepted. On his 92nd birthday, in 1985 Dart remarked: 'I knew people wouldn't believe me. I wasn't in a hurry'.[3]

In the 1960s and 1970s, many other important discoveries in the Olduvai Gorge and Great Rift Valley were made by Louis and Mary Leakey and their son Richard. The finding of Lucy by Donald Johanson and the oldest hominid *Ar. Ramidus* at 4.4 million years before present (mybp) by Tim White in Ethiopia helped to establish the lineage of early hominins. The debate at that time was mainly about the timing of the split between apes and humans, which had been estimated at around 30–40 million years ago, and it wasn't until DNA analysis was available that the date was brought much closer to 7–8 mybp.[4]

The other controversy was about which came first: a big brain or bipedality. This controversy was settled conclusively with the finding of Lucy and the Laetoli footprints, which confirmed that these early *Australopithecines* were bipedal but still had the brain of similar size to chimpanzees.

However, the important questions about the lifestyle and habitat in which these early hominins lived and why they, unlike any other terrestrial primate or mammal, became bipedal was largely overlooked. It was assumed that they had simply come down from the trees onto the savannah and acquired greater intelligence and manual dexterity, learning to make weapons and tools, and that it was easier for them to stand on two feet in order to see further on the savannah and to have both hands free.

A great deal has happened in the last 40–50 years to consolidate our knowledge about early hominin evolution, but still there are many who find it difficult to accept overwhelming new scientific evidence about the validity of the waterside theory of evolution. In *The Descent of Man*, Darwin had concluded: 'I can see no reason why it should not have been advantageous to the progenitors of man to have become more and more erect and bipedal'.[5] But then Darwin had not realised that becoming bipedal preceded a bigger brain and higher intelligence. Nevertheless, Darwin's concept of early humans becoming more bipedal on the savannah has remained entrenched in archaeological teaching and in the public perception.

ADAPTATIONS TO BIPEDALISM

Walking on two legs is something that adult humans take for granted, although as babies it is certainly not a natural mode of locomotion, and learning to walk is not easy. It took most of us about one year of struggling to sit upright, to crawl and, after many months of encouragement, support and tumbles, to take our first few tentative steps. Many babies, however, are able to swim long before they are able to walk[6] and certainly this is true of people brought up in coastal areas.[7–9]

Bipedalism does come at a cost, and certainly our early hominin ancestors would not have chosen to 'go it alone', 'wobble around' and learn to stand on two legs instead of four unless it provided a significant survival advantage to them. After all, they were still an ordinary mammal, like any other ape in their family, and had no human aspiration or any intellectual motive other than a basic instinct for survival. It is difficult to conceive that they suddenly had the intelligence, inspiration or instinct to walk upright on the savannah and go hunting for game. Like many other terrestrial mammals, such as the meerkats living on open ground, they might stand on two legs to get a better view of their surroundings, but as soon as there is a threat from a predator, the quickest way for them to escape is to scamper away on all fours!

On the other hand, at a time when food was scarce, searching for food in the shallow waters of the rivers and lakes provided them with a relatively easy way of finding nourishment because it was plentiful and easily obtained. It was natural for them to extend their territory into slightly deeper water and to find other sources of aquatic food, which even a pregnant female could do.

There are many images of apes and monkeys wading into the water for brief periods (Figure 17.1), but when back on land they revert to their natural quadruped gait for locomotion. The proboscis monkey, living in the mangrove swamps of Borneo, is the only other ape who has been observed to walk bipedally for short periods when on dry land.

To switch to a permanent bipedal mode of locomotion on land is quite a different matter, as it introduces so many new mechanical forces on the body and spine that would have been so foreign and difficult for our ape ancestors. In order for them have a better chance of survival, they must have become accustomed to spending quite a lot of time in the water looking for and gathering food, perhaps using sticks (later sharpened) to impale fish, squids and other aquatic animals. Hunting, swimming and diving for aquatic and marine food would have gradually resulted in strengthening of the lower limbs so that when they were back on dry land, like the proboscis monkey, they were able to support their body and start to walk on two legs. We do not have direct evidence

FIGURE 17.1 Bonobo foraging in the shallow water for an aquatic meal. (Courtesy of Gudkov Andrey/Shutterstock.)

of this other than the finding of copious deposits of fossil fish and crustacea, together with bones of terrestrial animals, in association with hominin remains.

One of the advantages of having a leg at each corner and a horizontal mammalian spine is great stability, providing support for the internal organs which are suspended below the vertebrae. Elaine Morgan describes it appropriately as resembling a 'walking bridge'[3] which seems ideal for a terrestrial mammal and one that can adapt to various habitats on the ground or in the trees.

For the early *Australopithecines*, however, there must have been a very good reason to abandon this tried-and-tested mode of locomotion and to gradually assume a more upright gait. It must have been advantageous, providing them with a better chance of survival which we assume was because they had to find new sources of food in the water. Swimming and diving to discover a plentiful supply of food was clearly an advantage to these primitive ape/hominins and a prime survival motivator for establishing and evolving major anatomical and physiological changes.

I do not think that any of these unique changes could have evolved without a major shift in the habitat of our ancestral hominin family from land to a water environment. Problems of weight distribution would not have been an issue if they were spending more and more time in the water, where there were effectively minimal gravity constraints. Buoyancy in the water would support the internal organs and allow gradual rotation of the pelvis to produce a bipedal and more streamlined mammalian shape, similar to transformations in the cetacean family. Early hominins found that they could utilise the benefits of becoming more upright to free their hands and upper limbs to carry objects and food, to make objects and weapons, and to pursue other domestic purposes.

LUMBAR DISC AND SCIATIC PROBLEMS

Several skeletal adaptations had to evolve over a period of time, particularly in the spine and pelvis, to support the internal organs. Apes were used to swinging in the trees, suspended by their arms much of the time. The vertical pressure on the spine would be minimal, with the weight of the body stretching the spine, tending to pull the vertebrae apart. Between the individual vertebrae are fibrous discs which act rather like strong elastic pivots, cushions or washers, providing stability and allowing a relatively wide range of movement in all directions in the lower back and in the neck. They also act like shock absorbers to absorb any excessive strain or to minimise the damage caused by an acute injury. In the thoracic part of the spine, the ribs are attached to each vertebra posteriorly and to the breast bone or sternum at the front of the chest.

This part of the spine has a more fixed and stable orientation and provides a protective cage for the heart and lungs. Movement of the chest is by vertical rotation of the ribs, pivoting on the vertebrae to allow expansion of the lungs.

For brachiating apes the pressure on the inter-vertebral discs would be minimal. However, one of the most conspicuous differences in the human spine is the progressive increase in size of the individual vertebrae from the neck down to the coccyx. The lumbar vertebrae at the bottom of the spine have to be much stronger to effectively carry the vertical weight of the body. They are also thicker and much more sturdy, and the increased load puts enormous pressure on the lumbar discs, as most of us are aware.

About 80% of adults experience low back pain at some point in their lives. In a recent large survey, the majority of adults had experienced some form of back pain in the preceding 3 months. It is the most common cause of job-related disability, and the magnitude of the burden from low back pain has grown worse in recent years. In 1990, a study showed that low back pain ranked sixth among the most debilitating conditions in the United States, but in 2010, low back pain jumped to third place, with only ischemic heart disease and chronic obstructive airway disease (COAD) ranking higher.[10]

In the United Kingdom, in 2016, an estimated 137 million working days were lost due to sickness or injury. This was equivalent to 4.3 days per worker, the lowest recorded rate since the series began in 1993, when the number was 7.2 days per worker. Minor illnesses (such as coughs and colds) accounted for the most days lost due to sickness in 2016, with 34 million days lost (24.8% of the total days lost to sickness), but musculoskeletal problems (including back pain and neck and upper limb problems) accounted for 30.8 million days (22.4%).[11]

Most lower back pain is acute and tends to be self-limiting after a few days or weeks. The symptoms may vary from an acute sharp incapacitating pain to a dull constant ache which is often worse on standing up. Degenerative bony changes and a sedentary lifestyle are important predisposing factors for low back problems, which affect men and women equally, particularly when a regular daily routine without much walking or exercise is punctuated by sudden bursts of strenuous activity. It may take the form of a local strain of one of the strong ligaments that support the spine from lifting a heavy object or from a sudden twisting injury or accident.

Other back problems may be due to herniation of one of the discs, most commonly the disc between the two lowest lumbar vertebrae. Sciatic pain is also a common problem due to compression of one of the roots of the large sciatic nerve which emerge from the intervertebral spaces or foramina of the lower lumbar and sacral vertebrae, just adjacent to the discs. If one of the discs prolapses and presses on the nerve root, the resulting sciatic pain is characterised by aching or shooting pains down the leg, numbness or paraesthesia (tingling, or 'pins and needles').

VERTIGO, NECK PAIN AND WHY GIRAFFES DON'T GET DIZZY

Cervical spine problems can also be a cause of symptoms in the neck, upper chest and arms. The cervical discs are not under as much vertical pressure as the lumbar spine because they only have the weight of the skull to support. However, there are two common ailments which affect the cervical spine, mainly due to degenerative changes of cervical spondylosis. There may also be damage from 'whiplash' injuries. Symptoms may include pain in the neck or upper chest from compression of the cervical sensory nerve roots as they emerge from the inter-vertebral foramina. They are also a common cause of headache affecting the back of the head, and the pain may also radiate around the shoulders and down the arms.

One other neck problem (discussed in more detail in Chapter 13) is the question of vertigo. I do see a lot of patients complaining of dizziness, and it is very important to find out what type it is, whether it is rotational or 'spinning' which may persist for some time, or whether it is a momentary episode of unsteadiness or feeling faint. The former is likely to be due to an inner ear problem affecting the semi-circular canals which register movements of the head. The latter unsteady episodes are more likely to be caused by brief reduction in the blood supply to the cerebellum.

The blood supply to the posterior 'primitive' part of the brain and cerebellum which controls balance is separate from the main blood supply to the brain through the carotid artery which one can feel pulsating on each side of the neck. The much smaller vertebral artery which supplies the cerebellum, as its name implies, runs up through small foramina in the cervical vertebrae (Figure 13.2). Because of the artery's smaller diameter, the posterior part of the brain is much more sensitive to a fall in blood pressure, for example, when getting up quickly from a sitting position or jumping out of bed too enthusiastically, when the cardiac output may fall by 20%. It can also happen when standing for long periods on a hot day, when the veins under the skin dilate and the volume of blood going back to the heart and pumped out again is again reduced.

The vertebral artery blood flow can also be reduced if there is any compression of the artery as it travels up the neck through the vertebral canals. This can happen with degenerative or arthritic changes in the cervical vertebrae, when extra thickening of the bone (osteophytes) can narrow the small canals. Typically this can be made worse on turning the head or looking upwards, when the artery gets 'kinked' with a sudden drop in blood flow. These problems are also more noticeable for someone who is on treatment for hypertension because these drugs lower the blood pressure and can lead to episodes of 'postural hypotension' when standing up or walking.

All of these problems with the vertebral artery and blood supply to the cerebellum seem to be unique to humans, simply because we stand upright and the head is no longer at the same level as the heart, as it is in most other terrestrial mammals, except, of course, the giraffe. Scientists have often wondered how a giraffe can lift its head from ground level within a few seconds to 15 feet above its heart and not feel faint, and also how it can lower its head to drink without suffering a catastrophic increase in blood pressure in the brain.

For a long time, it was thought that a large network of blood vessels at the base of the brain, called the carotid rete, helped the giraffe to avoid these two problems. Recent research has shown, however, that there is a drop in pressure across the carotid rete of only 1.5 mmHg, which is minimal considering that the typical blood pressure in a giraffe's head when it is drinking is around 330 mmHg.[12] There is still uncertainty about the answer to this question, but the rete might act as a sponge to reduce major pressure changes. The giraffe has a small head and brain and it has a really hefty 26-pound heart, much bigger than a human's. When its head is raised to reach up to a nourishing bit of foliage, it is possible that its heart is powerful enough to provide the necessary blood flow to the brain.

BLOOD PRESSURE AND SALT REGULATION

In terrestrial mammals, the control of blood pressure to maintain adequate perfusion of different parts of the body at all times, whether the animal is sleeping, standing or exerting itself in hunting or fleeing from a predator, is regulated partly by pressure monitors in the carotid artery in the neck but also through release of hormones which respond to emergency situations. The pressure monitors, or baroreceptors, are in the carotid sinus, situated in the wall of the internal carotid artery just above the bifurcation of the common carotid artery (Figure 13.2). In quadruped mammals, this is more or less at the level of the heart, but in hominins and humans, it is situated well above the heart, which has to be taken into account in controlling blood pressure.

Three main hormones help to regulate blood pressure, and they are produced in the adrenal glands which lie just above each kidney. They respond to changes in blood pressure detected by the carotid sinus. Adrenalin is not vital for life, but it plays an important role in the fight-or-flight emergency response by increasing blood flow to muscles and output of the heart, also causing pupil dilation so that our visual acuity is sharpened, and raising blood sugar to provide instant energy.

The main hormone essential for control of blood pressure is aldosterone, which responds to several different situations of anxiety and stress including surgery and loss of blood volume from haemorrhage. It also responds to an apparent loss of blood volume when humans stand up, when blood pools in the lower limbs causing a 20% drop in blood pressure. This drop is detected by the baroreceptors, which can't differentiate this from haemorrhage. Blood pressure and blood volume are also closely associated with salt balance, and aldosterone is vital in responding to salt deficiency in the body and conserving sodium by reducing its excretion by the kidneys. The sodium and water levels in our bodies are constantly changing as a result of dietary intake and sweating and have to be carefully balanced. Whenever there is an increase in salt levels in the blood, the kidneys reabsorb less sodium and more is excreted in the urine. On average, our kidneys can filter between 3.2 and 3.6 pounds of salt each day, which is about 150 times the amount of salt we ingest daily.[13]

Hydrocortisone (cortisol) is also vital and helps to regulate blood pressure and cardiovascular function, but it is mainly concerned with regulating how the body converts fats, proteins and carbohydrates into energy. In times of stress, cortisol also seems to be involved with release of sodium from stores in the sweat glands in our skin.

Corticosterone is another vital hormone produced by the adrenal glands which helps to regulate the immune response and suppress inflammatory reactions. Evolutionary changes in production of adrenal hormones have been essential for adaptation from aquatic to land-based environments. Lower vertebrates living in salty environments produce cortisol and corticosterone, whereas non-aquatic, terrestrial animals evolved to produce corticosterone and aldosterone.[14] Humans, on the other hand, then evolved to produce cortisol and aldosterone.

Contrary to advice about salt-restricted diets, it is essential to get enough daily salts and iodine to maintain vital body functions and avoid deficiencies caused by sweating, diarrhoea and vomiting to create the right fluid-sodium balance. Control of blood pressure in mammals is therefore a complex process combining pressure monitors, release of various hormones, and careful maintenance of salt and fluid balance. In humans, the control of blood pressure is much more difficult not only because of the relative positions of the heart and baroreceptors but also because of the extra pressure needed to pump blood from the lower limbs and the effects of gravity.

Other terrestrial mammals may suffer from high blood pressure, but the cause is usually due to some underlying hormonal disorder or tumour. High blood pressure in dogs is often associated with Cushing's disease or hyperadrenocorticism when a disorder, usually a benign tumour, allows an excess of cortisone to be released into the bloodstream. Cats often develop high blood pressure as a result of having hyperthyroidism caused by an over-production of thyroxine, a thyroid hormone that results in increased metabolism or in kidney disease.

HERNIAS, HAEMORRHOIDS AND PROLAPSES

When early hominins were beginning to walk bipedally on land, the weight of the abdominal contents would no longer be supported by the horizontal spine but exerted more and more pressure on the pelvis, which began to widen like a bowl to help distribute the additional load. We have seen how a compromise had to be reached between a wide pelvis, to accommodate this extra weight and allow a bigger birth canal, and yet to maintain narrow hips to maximise bipedal mobility and streamlining for walking, running and swimming. The extra intra-abdominal pressure caused by our vertical gait results in additional forces which are contained in the bony part of the pelvis but exert a strain on any weak spot, membranous part or opening in the pelvic floor.

In the lower anterior wall of the human male's abdomen in the groin area, on each side, the testes descend through a narrow opening and canal, called the inguinal canal, down into the scrotum, because sperm are much more efficiently produced and stored at a cooler temperature than that in the abdomen. This narrow opening and canal is a potential site of weakness and small bits of intestine can be forced into the canal, causing an inguinal hernia. These swellings are typically more noticeable when standing up or coughing and can be quite painful, but they can be repaired quite easily and the weak area can be reinforced with a strong artificial membrane. If an inguinal hernia is not recognised or repaired, there is a danger that the bit of intestine can become twisted, or 'strangulated', and become gangrenous, which is a dangerous situation and results in an emergency.

Prolapse can occur in women when one or more of the pelvic organs, the uterus, bowel or bladder, protrudes into the vagina. It is caused by weakening of tissues that support the pelvic organs. Although there is rarely a single cause, the risk of developing pelvic organ prolapse can be increased in older women, after a previous difficult or prolonged childbirth, in post-menopausal women with low levels of the hormone oestrogen, being overweight or having fibroids in the uterus. It can also be more likely to occur after previous pelvic surgery such as hysterectomy or bladder repair, or with chronic coughs, repeated heavy lifting and straining because of constipation.

Haemorrhoids or piles are also unique problems in humans and are due to dilatation and sometimes prolapse of dilated blood vessels around the rectum and anus from increased pressure in the veins in the vertical pelvis. They may be quite asymptomatic but can be painful and can cause bleeding. The predisposing factors are similar to those of prolapse, but they can affect men and women and are more likely to be caused by excessive straining from constipation or during pregnancy, being overweight, chronic diarrhoea and chronic coughing in smokers.

VARICOSE VEINS

In primates and most terrestrial mammals, the blood circulation in the body is more or less on a horizontal plane, apart from the legs where the venous blood going back to the heart needs to travel vertically upwards against gravity. In order for this to be achieved, the veins in the legs have one-way valves which allow the blood to flow in only one direction (back to the heart). When muscles in the legs contract, they squeeze the veins adjacent to the muscles, which forces the blood to circulate, but it can only go in one direction, back to the heart.

In the 1600s, William Harvey first performed a simple experiment showing that emptying and compressing a vein under the skin of the forearm in two different places a couple of inches apart will cause the vein to collapse. Releasing pressure on one (proximal) end will still leave the vein empty, but releasing pressure at the distal end (furthest from the heart) allows the vein to fill up again, showing that venous blood only flows in one direction, back to the heart.

When hominins stood upright, their upright position put extra pressure on the veins of the lower limbs, which made it much more difficult to force the blood upwards even further to the heart against gravity. About 90% of the blood returning up the legs goes via the deeper veins which are surrounded by muscle and are therefore not much of a problem. Varicose veins are more common in women than in men, and are linked with heredity.[15] Other related factors are age, pregnancy, obesity, menopause, prolonged standing, leg injury and abdominal straining. Sometimes pelvic vein compression and reflux can predispose one to varicose veins. When the pairs of flap valves in the superficial veins in the leg get worn down with excessive

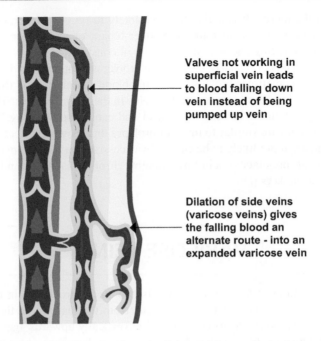

Valves not working in superficial vein leads to blood falling down vein instead of being pumped up vein

Dilation of side veins (varicose veins) gives the falling blood an alternate route - into an expanded varicose vein

FIGURE 17.2   Varicose veins – the superficial and deep veins in the leg. (Courtesy of the author.)

back pressure, the valves become incompetent and don't close off properly, which causes the veins to become dilated and twisted (Figure 17.2). Varicose veins can become unsightly and painful and often require treatment which has traditionally been surgical, but other endothermal or laser treatments have been advocated.

THE MEDICAL CONSEQUENCES OF OUR BIPEDAL HERITAGE

It seems that everything comes at a price and that the final benefit of the supreme outcome of survival has to be weighed against the risks and side effects of whatever changes have taken place. When doctors are faced with patients who have cancer or a life-threatening disease and may eventually die from that condition, they have to carefully consider what options of treatment are available and the chances of success. For many cancers and cardiovascular diseases, the treatments now available have a very good chance of success. Sometimes, however, the consequences of radiation and chemotherapy may have long-term side effects which might affect the quality of life. With our continued research into the causes of cancer and other life-threatening diseases, we hope that in time we shall be able to cure and eradicate these conditions, as we have done in the past with infectious diseases.

Various conditions have become medical problems in humans because of our upright stance, and they are not seen in other terrestrial quadruped animals. These conditions must surely have had their casualties over the past few million years, but they have not been sufficiently substantial to eradicate the benefits of bipedalism and our waterside existence. Hominins have managed to survive by adapting to various epigenetic influences in their environment and have eventually, thanks to the cognitive revolution, had the benefit of increased intelligence and resourcefulness. This is evident in *Homo sapiens*, who have worked to understand and treat these conditions, which otherwise would have caused extinction of our species long ago.

We Are What We Eat

18

Humans were at the water's edge, foraging in relatively shallow waters, as well as spending time on land, for things like shellfish, aquatic plants, crabs, things that are not going to run away, or swim away from you. You don't have to be fast, you just need to be able to hold your breath for a minute or so, go under the water for a few metres, pick up a few shellfish and you have got yourself a meal, something that pregnant women can do.
Sir David Attenborough (September 2016)

Food is essential for survival and has been one of the most important factors in human evolution and demographic growth. Roughly 10,000 years ago, the remarkable growth and development of increasingly complex social structures followed the shift in the provision and management of resources from hunting and gathering to agriculture and stock breeding. Food preservation and cooking are not only important indicators of culture and civilisations but have also been major determinants in evolutionary history, human survival and morphology.

It has long been assumed that early hominins evolved from their primate ancestors, leaving the forests to become adapted to life in an open, arid, savannah habitat. It is suggested that they gradually assumed a more upright gait and bipedal locomotion to adopt a hunter-gatherer lifestyle with long-distance running in order to hunt and scavenge large mammals.[1,2] This model pre-dates Darwin's theory of natural selection[3] and has never really been subjected to rigorous scientific scrutiny.[4] The idea was formulated well before the vast majority of fossil and other evidence of human evolution was discovered, as *post hoc ergo propter hoc* ('after that and therefore as a result of that'), on the assumption that they just came down from the trees onto the ground.

The idea that early hominins may have hunted large mammals by running them to exhaustion makes the huge assumption that they suddenly became transformed from a simple member of the chimpanzee family to one that had the ability not only to stand, walk and run proficiently on two legs but also to outdistance fast-running game on an open savannah.[5] There is no scientific evidence of parallel development in any other terrestrial mammal to stand upright and walk and run as a preferred mode of locomotion.[4] Such a dramatic unique evolutionary change in a mammal on the savannah would have taken hundreds of thousands of years to perfect, and it is difficult to comprehend how a chimpanzee learning to walk and run on two legs could have survived predators or found enough food to survive in dry, open landscapes.[5,6]

The Taung child was discovered in the 1920s by Dart[7] in an arid habitat that he assumed had changed very little since the child died there. But we now know that the landscape was once forested and much wetter.[4,8] It seems far more likely that *Australopithecines* and early hominin populations were living in waterside habitats and capable of bipedal wading, swimming, underwater foraging and diving, which is how they obtained their main sustenance.[9] During this long period of time, when they were gradually evolving a rotated pelvis and upright gait, they slowly gained strength in their lower limbs and perfected the ability to walk bipedally and later to run on land.

Only later were they able to extend their hunting techniques and strategies to include chasing, catching and acquiring meat from larger land-based animals. Even then, with the advent of the genus *Homo*, it is questionable whether they were able to gain enough sustenance with meat in terrestrial savannah settings.[10–13] They likely would have needed the additional vital nutrition from easily obtainable aquatic sources.

Although the occasional use of fire was identified from about 800,000 years ago, the regular daily use of fire and advent of cooking food by *Homo erectus*, Neanderthals and other hominids around 300,000 years ago was a major advance which allowed them to eat and digest otherwise rather tough meat and vegetables much more easily. They were also able to widen their diet to include previously indigestible food. It allowed them to smoke and cure meat and fish, preserving them for times when food was scarce, especially in the winter months. Fire was therefore one of the most important evolutionary and survival factors which widened the gulf between humans and other animals.

EARLY HOMININS AND THE CRUCIAL ROLE OF MARINE AND LACUSTRINE FOODS

There is increasing scientific evidence that *Australopithecines* and other early hominins evolved in a waterside environment, initially foraging in rivers, estuaries, salt water lakes of the Great Rift Valley and other wetlands and acquiring skills of swimming and diving which became adapted for bipedal walking and running on land.[9] It is suggested that the combination of anatomical and palaeo-ecological data are incompatible with the savannah theory.

From an anatomical point of view, the bipedal *Homo erectus*, *Homo ergaster* and *Homo heidelbergensis* were completely unsuited for long-distance running. In particular, their skeletons revealed pachyosteosclerosis, which is a common feature of all 'erectine grade' fossils, with thickening of certain bones, a narrower marrow cavity, thicker cortices and denser bone.[4] These hominins were also more broadly built than *Homo sapiens*, with a wider pelvis; shorter stature; a thicker occipital area of the skull, indicating more powerful neck muscles; and thickening and flattening of the femur, which Kennedy had described in *Homo erectus* as remarkably significant[14] because it was unique for a terrestrial mammal. The only other creatures to show this feature of pachyosteosclerosis are slow-moving marine mammals such as the dugong (or sea cow) and walruses that forage in relatively shallow waters for sea grasses and immobile or sessile foods.

In modern times, long-distance runners are typically slim and have light bones, as do fast-moving aquatic mammals such as dolphins. Heavy bones would be a distinct disadvantage for a fast-running hunter-gatherer on land, but there does appear to be a clear correlation in extant animals and the fossil record between aquatic foraging for relatively immobile food and pachyosteoscerosis.[4] Other anatomical features of hominins which suggest a waterside habitat with foraging, swimming and diving are loss of body hair, subcutaneous fat, voluntary breath control, the newly described unique mechanism which caused descent of the larynx from diving, a streamlined body and an elongated nasal shape and nasal valve to protect the airway when swimming.

From a nutritional and biological point of view, the essential ingredient which allowed hominins to acquire a large brain and higher intelligence, like other aquatic mammals, was the abundant availability of the two phospholipids docosahexaenoic acid (DHA) and arachidonic acid, as well as iodine and other essential salts, which were not so readily available on the savannah. DHA is not synthesised by terrestrial plants, and although some large ruminants have high levels of DHA in their brain tissue, the meat and bone marrow contain very little.[9]

The consumption of DHA is particularly essential during pregnancy and during the first few months of life because of the dramatic surge in brain growth. There is substantial evidence that long-chain polyunsaturated fatty acid (LC-PUFA) deficiencies of these phospholipids is associated with attention-deficit/hyperactivity disorder (ADHD), dyslexia, senile dementia, depression, bipolar disorder, anxiety, schizophrenia and other neurological and psychiatric disorders.[9,15–22] Animal research has also shown that these problems increase in severity as successive generations continue to be deprived of LC-PUFA.[9,23–25]

Marine and freshwater plankton and algae provide the richest source of DHA, either by direct consumption or through a diet including fish and crustacea and other algae-eating amphibians, reptiles, birds and marine mammals. Stewart and Cunnane have made a detailed review[26] of numerous early hominin archaeological sites, including the Olduvai Gorge,[27] Middle Awash Valley[28] and Turkana Basin[29] where fish remains are common.

Munro has carried out a large palaeo-eceological survey to investigate the presence of molluscs and divided the associated hominin and non-hominin sites broadly into terrestrial, semiaquatic and aquatic.[30] All of the 'erectine grade' hominin sites revealed aquatic and marine molluscs, indicating the presence of water, which was not the case for the non-hominin or non-erectine grade sites. He also pointed out that because most of the Pleistocene period (2.6 million to 11,700 years ago) was during the ice ages, sea levels were lower then. Hominin coastal sites from that period are therefore now largely underwater, and few hominin remains have been found. However, there is ample evidence that large bivalve mussels and oysters were consumed in many different sites from Africa, the Middle East, Italy, Pakefield in the United Kingdom, Indonesia and Australia, often in association with stone tools, and they continue to be an important source of food in many modern populations.

The anatomical and palaeo-ecological data seem to be inconsistent with the long-distance running hunter-gatherer savannah model of early hominin evolution, but they are perfectly compatible with a littoral habitat and lifestyle that includes foraging, swimming and diving for slow-moving and sessile aquatic and marine food. This would provide early humans with the phospholipids essential for evolution of a larger brain and higher intelligence, culminating in the emergence of *Homo sapiens*.

THE COGNITIVE REVOLUTION

By the time that *Homo sapiens* first appeared in South Africa around 200,000 years ago, hominins were becoming much more domesticated in their waterside habitats. Fire was routinely used for cooking, which meant that, unlike chimpanzees, they no longer had to spend 5 hours a day chewing raw food. Their gastrointestinal tract had become shorter and more efficient in digesting food, and the resulting shift of circulating blood to the increasing requirements of the brain allowed rapid progress in more complex neurological circuits and intellectual sophistication. Most scientists agree that by 150,000 years ago, our ancestors in Africa were very like us in appearance and were evolving a much more independent lifestyle based on a cohesive and co-operative community.

For whatever reason, the migration of early *Homo* out of East Africa began around 70,000 years ago via coastal routes to the Arabian Peninsula and beyond to Europe and Asia. From DNA analysis, we know that they lived and interbred with Neanderthals in Europe and the Denisovans in Indonesia before reaching Australia by 45,000 to 50,000 years ago. There is strong evidence that their coastal way of life remained the same, as it did for the Australian Aborigines and for many coastal populations to the present day.

Mount Carmel in the Middle East has been one of the most fascinating archaeological sites and has provided an archive of early human life in southwest Asia. It contains cultural deposits representing at least 500,000 years of human evolution demonstrating the unique existence of both Neanderthals and early anatomically modern humans (AMHs) within the same Middle Palaeolithic cultural framework, the Mousterian (160,000–40,000 years ago).

Evidence from numerous Natufian cave burials and early stone architecture represents the transition from a hunter-gathering lifestyle to agriculture and animal husbandry. As a result, the caves have become a key site of the chrono-stratigraphic framework for human evolution in general, and the prehistory of the Levant in particular (Figure 18.1).

Understanding the behavioural adaptations and subsistence strategies of Middle Palaeolithic humans is critical in the debate over the evolution and manifestations of modern human behaviour. The discovery of faunal remains from Misliya Cave, Mount Carmel, Israel (>200,000 years ago), allowed for detailed analyses of these early Middle Palaeolithic remains.[31]

The Misliya Cave faunal assemblage is overwhelmingly dominated by ungulate taxa. The most common prey species is the Mesopotamian fallow deer (*Dama mesopotamica*), followed closely by the mountain gazelle (*Gazella gazella*). Some auroch (*Bos primigenius*) remains are also present.

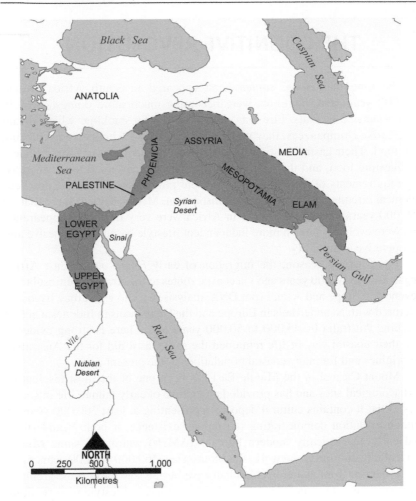

FIGURE 18.1   The Levant. (Courtesy of Norman Einstein.)

Small-game species are rare. The fallow deer mortality pattern is dominated by prime-aged individuals. A multivariate analysis demonstrates that:

1. The assemblage was created solely by humans occupying the cave and was primarily modified by their food-processing activities.
2. Gazelle carcasses were transported complete to the site, while fallow deer carcasses underwent some field butchery.[31]

The new zooarchaeological data from Misliya Cave, particularly the abundance of meat-bearing limb bones displaying filleting cut marks and the

FIGURE 18.2 Palaeolithic tool from Mount Carmel. (Courtesy of Strannik_fox/ Shutterstock.)

acquisition of prime-age prey, demonstrate that early Middle Palaeolithic people possessed developed hunting capabilities (Figure 18.2). Thus, modern large-game hunting, carcass transport, and meat-processing behaviours were already established in the Levant in the early Middle Palaeolithic, more than 200,000 years ago.[31]

AGRICULTURAL ORIGINS AND THE NEOLITHIC REVOLUTION

The beginning of agriculture around 12,500 years ago[32] has been regarded as one of the most important landmarks in the evolution and culture of *Homo sapiens*. The gradual transition from a waterside lifestyle of hunting and gathering, both on land and in the water, to a more settled one of agriculture and domestication allowed expansion of hominin communities and the population,[33] with increasing group co-operation and the development of social organisation and hierarchies. These more settled communities allowed humans to observe and experiment with plants to see how they grew and developed, leading to domestication of various types of plants.[34,35] It is claimed that the

Neolithic Revolution dramatically narrowed the dietary niche by decreasing the variety of available foods, with the shift to intensive agriculture creating a dramatic decline in human nutrition.[36]

Agriculture began independently in several centres in the New and Old Worlds, and at least 11 areas have been identified. Wild grains were collected from about 22,000 years ago at various sites, but from about 9,500 BC the eight Neolithic recognised founder crops were cultivated in the Levant, which is a large area in Southwest Asia bounded by the Mediterranean Sea in the west, the Arabian desert in the south, Mesopotamia in the east and the Taurus mountains in the north (Figure 18.1). Rice and then soy were domesticated in China between 11,500 and 6,500 BC. Sugarcane and some root vegetables were domesticated in New Guinea around 7,000 BC.

In the Andes in South America, the potato, along with beans, cocoa, llamas and guinea pigs were domesticated between 8,000 and 5,000 BC, and in the same period bananas were cultivated in Papua, New Guinea. Other animals such as pigs were also domesticated in Mesopotamia around 11,000 BC, followed by sheep between 11,000 and 9,000 BC; shortly afterwards cattle were domesticated from the wild aurochs in areas of Turkey and Pakistan around 8,500 BC. The use of cotton was introduced in Peru much later, by 3,500 BC, and camels were domesticated around 3,000 BC.

Agriculture has therefore had an important and long-lasting impact on human demography, genetic variation and culture and on the earth's environments. It has facilitated a near global shift to more sedentary lifestyles and a vast increase in human population levels. The Neolithic Revolution resulted in much more than adoption of a limited way of cultivating and producing food. It transformed the small and mobile groups of semi-aquatic hunter-gatherers into non-nomadic societies and communities based in small villages and towns.

Many of the early centres of urbanisation and cultural developments were based around the coastal areas of the Levant and its three main rivers, the Nile in Egypt and the Tigris and Euphrates in Mesopotamia; these areas became known as the Fertile Crescent. Waterside communities became larger with increased organisation into states, and although humans retained their strong affinity for the sea and water, their domestic lifestyle had to be based on land.

Deforestation and irrigation of the land became more important for specialized food cultivation which allowed the production of surplus food that could then be exported as trade to other countries and communities. Diversification and development of other material technologies such as pottery, ceramics and metal followed with the advent of the Bronze Age and the Iron Age. By the Middle Ages, new and improved techniques transformed agriculture. In the fifteenth century, contact with the New World brought exchanges of maize, potatoes and manioc from the Americas. Old World crops such as wheat, barley, rice and turnips as well as livestock, including horses, cattle, sheep and goats, were transported to the New World. Further advances were made in the seventeenth through nineteenth centuries with the British agricultural and industrial revolutions and crop rotation.

THE GLOBAL FOOD CRISIS

In more modern times, as far as agriculture is concerned, during the last century, increased mechanisation and the use of synthetic fertilisers, pesticides and selective breeding have transformed agriculture in the developed nations, greatly increasing crop yields, but it has resulted in social, political and environmental issues. Organic farming became more popular as an alternative to synthetic pesticides in developed countries, but famines continued to affect many areas around the globe through adverse climatic events, wars and crop failures. As a result, many millions of people have died. Initiatives, such as the Green Revolution, led by Norman Borlaug in the 1970s, increased agricultural production around the world, and he is credited with saving over a billion people from starvation (he received the Nobel Prize in 1970).[37,38]

However, the critical situation of world poverty and starvation needs to be addressed urgently. The 2018 Global Report on Food Crises provides the latest estimates of severe hunger in the world. An estimated 124 million people in 51 countries are currently facing a crisis in food insecurity or worse. Conflict and insecurity continue to be the primary drivers of food insecurity in 18 countries, where almost 74 million food-insecure people remain in need of urgent assistance.

The report from 2017 identified 108 million people in crisis of food security or worse across 48 countries. A comparison of the 45 countries included in both editions of the report reveals an increase of 11 million people – an 11% rise – in the number of food-insecure people around the world who require urgent humanitarian action.[39]

An Incredible Journey

19

When dawn breaks tomorrow on this small planet, seven types of great ape will get up out of their comfortable beds, stretch their limbs and start contemplating the first meal of the day. Six of these apes are on the verge of extinction, one is an unparalleled success. The one that is on the verge of exterminating them is the naked ape, the animal usually referred to as a human being.
Desmond Morris
Planet Ape (2009)

In the early years of the twenty-first century, there were 6.6 billion naked apes occupying almost every corner of this precious planet. Forty years ago, there were only 3.3 billion of them, and so their numbers have doubled since then. Most species have ways of controlling their numbers so that their populations stabilise and they do not outstrip their resources, but naked apes lack these controls because, through increasing intelligence, they have learned how to harness and change the environment to suit their needs so that they can attempt to control their destiny.

Attempting to write an account of human evolution and to describe how we have amazingly succeeded in evolving from a simple arboreal quadruped ape into a sophisticated intelligent human is rather like finding some old family photographs and trying to piece together from a few snapshots a picture of what their lives must have been like, gathering information from as many other sources as possible. The only difference is that this story goes back at least 6 million years. The hard evidence in terms of fossil finds is scarce and a lot of the other information is just circumstantial. Recent advances in genetics and DNA analysis are giving us a great deal more detail, however, about the historical composition of our genomes and exactly who were our true ancestors.

One of the problems has been the tendency for scientists to base too much reliance on assuming that just because the quadruped apes and modern humans are both based on *terra firma*, the sequence of evolutionary events had to be based on land and that no other intermediate phase or influence was involved. There has also been a bias to make judgements founded on modern contemporary thoughts, ideas and intelligence rather than to consider early hominin evolutionary changes in the context of that particular time. We were, after all, just a branch of the ape family, and we still are animals like any other creature. One important aspect of evolution is that, like other animals, our anatomy and physiological and metabolic processes are exactly the same as any other animal. They evolve and adapt to changing environments and external influences, the epigenetic forces that Darwin felt were so important. The intellectual influences came very much later, only in the last 200,000 years.

In the early days of the *Australopithecines*, our ancestors were physically and intellectually just like any other chimpanzee, fighting for survival in the wake of the late Miocene drought when food was scarce. Unlike the gorillas, chimpanzees, bonobos and other primates, who stayed in the same ecological niche in the forests where they have remained the same until the present day, a few of them started foraging for food in the rivers and lakes, some of which, around the Olduvai Gorge and Great Rift Valley, were salt lakes.

One can easily imagine that the aquatic plants, catfish, molluscs, mussels and other crustaceans were readily available to any quadruped ape wading into the water and that gradual extension of their territory into deeper water in search of a more varied diet would have been a natural and simple progressive step. Gradually adopting a more upright stance in the water to keep their airway safe, the buoyancy supporting their legs, spine and body, one can again imagine these apes slowly adapting to this habitat, being able to escape into the water or trees from terrestrial predators and likewise avoiding crocodiles and other aquatic hunters by going back on land. Becoming more accustomed to this semi-aquatic lifestyle, our ancestral apes gradually evolved all of the other unique aquatic characteristics of hairlessness, subcutaneous fat for insulation and streamlining, adaptations in the kidneys and thermoregulation seen in other littoral mammals but totally unheard of in terrestrial mammals.

By 3–4 million years ago, they were now able to walk on land, freeing their hands to develop increasing manual dexterity, which gave them a distinct advantage over their primate cousins. Meanwhile, their aquatic diet gave them the essential phospholipids which were vital in allowing progressive neurological development and sophistication in their brain.

This evolutionary scenario seems very much more plausible than the alternative traditional savannah theory, which has no comparable example in the evolution of any other terrestrial species. Not one convincing argument has been proposed in the past 150 years to explain why one branch of the ape family, let alone any other land-based animal, was driven by a survival instinct to make the gradual change to standing upright, which must have taken place over at least several hundred thousand years. During this time, anatomical and physiological adaptations would have had to be made to allow rotation of the pelvis, sufficient strengthening of the lower limbs to take the extra weight when mobile, and learning to balance on two legs instead of four, especially when moving and then running.

The simple reason why no other terrestrial animal has done this is demonstrated by other land-based mammals such as the meerkat, chimpanzee, bonobo and proboscis monkey. All can stand on two legs but whenever there is danger and the need to move quickly, they can escape much more easily on all fours. In my opinion, any ancestral hominins persisting in their attempts to try and learn a bipedal mode of locomotion on the savannah would very soon fall prey to one of many predators looking for an easy meal. They would never have survived long enough on the savannah to develop the skills required to move and run on two legs. How could they hope to outrun other quadruped animals and game like antelopes when it would have taken them many thousands of years on the savannah just to learn to stand and move bipedally?

Quite apart from the explanation of bipedalism and the problem of surviving on the savannah looking for game, how does the savannah theory explain all of the other unique anatomical and physiological characteristics in hominins such as subcutaneous fat, hair loss, exostoses, a different thermoregulatory system, differences in the kidneys and salt excretion and the eventual growth of a big brain and higher intelligence? There is also the question of human speech. Not one credible or logical argument has been made by supporters of the savannah theory to explain how these could have developed in early hominins as terrestrial mammals.

THE ORIGIN OF SPEECH AND LANGUAGE

Articulated language is regarded as the one of the most fundamental features distinguishing humans from all other animals. It is important to recognise, however, the difference between the ability of humans to make continuous vocal sounds; second, the ability to make articulated intelligible speech; and third, the evolution of language. Other mammals are able to bark, roar or cry by temporarily lowering their larynx, but immediately the larynx returns to its original position, with the epiglottis and the soft palate adjacently interposed, to resume obligatory nasal breathing. This includes human babies who can breastfeed and breathe at the same time but are able to cry by temporarily lowering their larynx. The difference with children older than about 6 to 12 months and adult humans is that the larynx slowly grows into a permanently descended position, not seen in any other mammal, which allowed us to vocalise sounds freely at any time. The longer and more mobile tongue, together with use of the palate, teeth and lips, permitted a more versatile production of different sounds including vowels and consonants.

Ann Maclarnon is director of the Centre for Research in Evolutionary Anthropology at Roehampton University and has provided an excellent analysis of the anatomical and physiological basis of human speech production.[1] She acknowledges that our closest primate relatives demonstrate nothing close to our unique human form of communication and that the physical feature of descent of the larynx enabling voice may have developed as specific adaptations or exaptations. She explains that, although the vertical lengthening of the tongue base and separation of the soft palate and epiglottis commenced after weaning, the final ratio between the length of the oral tongue and vertical base of the tongue portion only reaches 1:1 from about 6–8 years of age.

She quotes Lieberman,[2] who suggested possible adaptational reasons for the lengthening of the tongue base and descent of the larynx, including intracranial temporal lobe increase for enhanced cognitive processing including language; increasing flexion of the skull base; increased meat consumption and technologically enhanced food processing including cooking, resulting in facial reduction; endurance running; building on obligate bipedalism, involving facial reduction for improved head stabilisation; direct selection for speech capabilities, driving a decrease in oral cavity length, involving facial reduction and/or skull base flexion, to produce a 1:1 ratio. However, I believe that these multiple suggested explanations are unphysiological and vague, without any rational scientific evidence.

Maclarnon suggests other possible causes of laryngeal descent but concludes that it may have been due to earlier evolutionary change. Many other authors have proposed different mechanisms for descent of the larynx but as previously explained, all of these theories have been based on the presumption that hominin humans were terrestrial mammals living on the savannah, which is why none of them are physiologically or anatomically possible. None of them had ever considered the alternative semi-aquatic explanation.

Two of the most important skills that early hominins were able to develop in the waterside environment were swimming and diving, learning to hold their breath for longer periods so that they could go down to the bed of the lake or sea in search of a more varied diet. The physiological changes in the heart, circulation and lungs associated with diving were something that continues to be seen in native coastal communities, even in children and babies before they can walk and in professional divers to this day. In my opinion, the negative intra-thoracic pressure causing downward traction of the trachea, larynx and base of the tongue during diving resulted in descent of the larynx, which evolved as a unique human characteristic. This caused separation of the soft palate and larynx and vertical elongation of the tongue to double its length, but in a vertical plane, producing the 1:1 ratio.

The evolution of speech was therefore very much a two-stage development in humans. The initial phase was the descent of the larynx as a result of diving and breath holding, with separation of the nasal airway and the pharynx, which presumably gradually developed over several hundred thousand years. This set the stage for the second phase in the course of the cognitive revolution during the last 200,000 years, when increased intelligence allowed the unique difference in the anatomy of the human vocal tract to be used for development of more meaningful sounds and communication.

It is difficult to be certain about the timing of speech evolution, and many theories have been proposed. In 2005 Deutscher stated: 'Failing the discovery of a camcorder left behind by careless aliens on a previous visit, it is thus difficult to see how the first emergence of speech in hominids can even be much more than the stuff of fantasy'.[3]

THE CHALLENGE OF MENTAL ILL HEALTH

Despite major advances in agriculture, communication, medicine, manufacturing, mechanisation and better education, the world is a place of great health and economic inequality, with poverty affecting nearly half the global population.[4,5] Global population eventually reached 1 billion in 1804, but it took only 123 years to reach 2 billion in 1927 and an additional 33 years more to reach 3 billion. Forty years later, in 2000 it had doubled to 6 billion, and it is expected to reach 8 billion by 2020.

One hundred years ago, tuberculosis, flu, gastrointestinal infections, pneumonia and diphtheria were the leading causes of death in the United States, but these have largely been eradicated due to antibiotics, vaccination, improvements in sanitation and public health, and more effective medical treatments. As the impact of these diseases has been reduced or eliminated, there has been a significant increase in mortality from heart disease, colon and breast cancer, obesity, type 2 diabetes, hypertension and suicide, despite major attempts at health education. Diabetes caused 1.6 million (2.8%) deaths globally in 2015, up from 1.0 million (1.8%) deaths in 2000. Deaths due to dementia more than doubled between 2000 and 2015, making it the seventh leading cause of deaths worldwide in 2015.[6]

In 2004, the European audit of health revealed that the cost of brain-related disorders had overtaken all other burdens of ill health at E386 billion for the 25 member states.[7] By 2010 the cost had risen to E789 billion. In the United Kingdom, the cost of mental ill health had risen from £77 billion in 2007 to £105 billion in 2011, more than the cost of heart disease and cancer combined. In 1972, Michael Crawford, professor in mental health research in the Centre for Psychiatry at Imperial College, London, predicted that there would be an increase in mental ill health and brain disorders[8] and there is now substantial evidence that deficiencies in long chain poly-unsaturated fatty acids (LCP-UFA), mainly derived from seafood, are linked to attention deficit hyperactivity disorder (ADHD), depression, bipolar (manic depressive) disorders, suicides, self-harm and anxiety disorders.[9]

Of additional concern is that animal research has shown that these problems of mental disorders increase in severity as successive generations continue to be deficient in phospholipids.[9,10] There was great wisdom in the age-old tradition of eating fish on Fridays, which sadly has not continued in modern times. Encouraging a regular dose of cod-liver oil, even if it was not so palatable, has also become a thing of the past.

Furthermore, fish oil supplements have provided benefits to psychiatric patients, particularly for major depression and feelings of self-harm.[9,10] Without an adequate supply of these brain-specific nutrients, stunting of brain development is inevitable since fetal, maternal, infant and pubertal malnutrition all affect the developing brain. The impacts are irreparable and permanent, promoting poverty, psychosis or depression for yet another generation.[11]

More than 450 million people around the world have mental, neurological or behavioural problems, yet the vast majority lack protection and appropriate treatment. Mental health legislation is missing or outdated in 64% of countries, and 30% lack a budget for mental health.[12]

HEALTH, POPULATION GROWTH, SOCIAL INEQUALITY AND POVERTY

Poverty, diminished food intake, uncontrolled birth rates and poor antenatal care are prime causes of social inequalities and underdevelopment worldwide. One of the major problems in health planning in many countries is the fact that most of the influences which may have highly significant implications for health – such as housing, agriculture, finance, social services, education, engineering and infrastructural developments – lie within government sectors that have no responsibility for health.[12] In the absence of stabilisation of population growth, the fight for improved education, health, development and elimination of poverty becomes much more difficult.

Over 50% of the world's population lives in the developing world, where health problems are most severe. Cancer mortality accounts for 76% of deaths in these countries, which have only 5% of the resources. The burden of disease is greatest in these regions of the world, notably Africa and Asia, where resources are few. In Malawi there is one doctor per 100,000 people and only one ear, nose and throat (ENT) surgeon for its 12 million people.[13] There are 30 million HIV patients in Africa and 11 million AIDS orphans. The incidence of tuberculosis is 900 per 100,000 population; most of them are multi-drug resistant and over 50% are HIV positive.

Survival of people with cancer in Africa, where 30 out of 52 countries have no radiotherapy facilities at all, is far worse than that in high-income countries. For example, the 5-year survival rate for women with breast cancer in Europe is 82%, whereas it is 46% in Uganda, a little less than 39% in Algeria and 12% in Gambia. In 2012 nearly 850,000 new cancers appeared in Africa, while in that same year, there were almost 600,000 deaths attributed to malignant disease. The mortality/incidence ratio, an expression of the efficacy of the cancer care system, was 72% compared with 44% in Europe that year.[14]

The global epidemic of obesity, second to cigarette smoking as the highest preventable cause of cancer, is itself a risk factor for several other serious health problems such as diabetes, hypertension, stroke and other non-communicable disorders. It has been the subject of many reviews by the World Health Organisation (WHO)[15] and Pan American Health Organisation (PAHO)[16] and other more recent publications.[17,18] In many affluent countries at least half of the population is overweight and one in five is obese. Rates in excess of 50% and up to 75% are found in some island countries in the Western Pacific and other islands in the Caribbean.

AN INCREDIBLE JOURNEY

It has been an incredible journey over the past 6 million years, which has seen the remarkable evolution of modern humans from a simple primate who, unlike their other primate cousins, chose a different semi-aquatic habitat and source of food, with the one intention of survival at a difficult time of food shortage. There was no inevitability about this process, and little did our early hominin ancestors realise what was to be the eventual outcome. But nor do we know what the future will hold, since evolution continues to progress.

However, unlike our early hominin ancestors who were just trying to survive in a hostile environment, the difference now is that we are in control of our own destiny, barring any extraterrestrial or worldwide disaster. I hope that this book has helped to give a logical explanation for some aspects of human evolution. One reason that there have been so many unproven theories proposed during the last 150 years about why we are so uniquely different from our ape cousins, including hairlessness, bipedalism, subcutaneous fat, big brains, different thermoregulation and voice, is that they have been based on the preconceived notion that we evolved as terrestrial savannah apes.

We have to thank Sir Alister Hardy and Elaine Morgan for sowing the seeds of doubt and for introducing the idea of an aquatic phase in our evolution. Thirty years ago, when I was trying to make sense of the expanded sinuses in humans, a little bone in the external ear canal and other unique anatomical and physiological features in humans, I tried to combine the knowledge from fossil evidence with basic medical principles. The words of the French Renaissance physician, Jean Fernel, from the beginning of the sixteenth century, seemed important: 'Anatomy is to physiology as geography is to history; it describes a theatre of events'. I had little idea that my search for answers would lead to a radical proposal that humans have evolved from our primate family as semiaquatic mammals and not as terrestrial mammals, as is generally believed.

Unlike our primate cousins, we have a waterside ape heritage that explains so much about our natural affinity to water and the sea. On September 14, 1962, at Newport, Rhode Island, on the occasion of the America's Cup, President John F. Kennedy said: 'I really don't know why it is that all of us are so committed to the sea, except I think it is because in addition to the fact that the sea changes and the light changes, and ships change, it is because we all came from the sea. And it is an interesting biological fact that all of us have, in our veins the exact same percentage of salt in our blood that exists in the ocean, and, therefore, we have salt in our blood, in our sweat, in our tears. We are tied to the ocean. And when we go back to the sea, whether it is to sail or to watch it, we are going back from whence we came'.

It is our responsibility to look after our world and our fellow human beings and safeguard the future for our children and generations to come. The search for an improved quality of life and the relevance of health is essential, not only for poor countries but for affluent ones as well.[12] If we are to do this successfully, it is essential that we grow to understand as much as we can about ourselves, our evolution and history, about the nature of life on this incredible planet earth and our place in it.

Glossary

AAT: aquatic ape theory. Introduced by Sir Alister Hardy in 1960 and later the subject of several books by Elaine Morgan, it proposed that early human evolution involved a significant semi-aquatic phase.

actinic keratosis: relating to radiation, such as ultraviolet radiation, which produces a chemical effect on the skin that becomes thickened. A precursor to development of skin cancers.

adipose: fat under the skin and surrounding major organs, providing stored energy, insulation and protection.

aero-digestive tract: the upper aero-digestive tract is part of the respiratory system and feeding channels including the mouth, the nose, the throat, the voice box and the vocal tract.

Afar region: a peninsular area in northeast Africa where early hominid fossils were found.

AMH: Anatomically Modern Humans. The first *H. sapiens* to appear in Europe.

apnoea: a temporary suspension or absence of breathing,

apocrine: describes glands in the skin that secrete an oily substance for lubrication of hairs and scent recognition. Some are specialized to produce wax in the ear and breast milk.

Ardepithecine: An extinct early hominid from 4-5 to 4.3 million years ago that was bipedal but had tree-climbing abilities.

Aurignacian technology: sophisticated tool and art making, developed during the 'Cognitive Revolution' by *H. sapiens*.

Australopithecines: the earliest bipedal hominins, including *A. afarensis*, *A. africanus*, *A. ardepithecus*, *A. bosei*, *A. giganticus* and *A. robustus*.

autonomic: describes functions of the nervous system not under voluntary control, e.g. the regulation of heartbeat or gland secretions. The functions are either sympathetic (associated with the fight or-flight response) or para-sympathetic, often with opposing functions (e.g. increasing or decreasing heart rate)

bipedal: describes an animal whose main mode of locomotion is with two legs or two feet.

Black Hole: a period of evolutionary time from 9 to 5 million years ago for which few pre-hominid primate fossils have been found.

bonobo: a rare black arboreal chimpanzee, native to west Africa, south of the Congo River.

bradycardia: slow heart rate, usually measured as fewer than 60 beats per minute in an adult human.

carcinogenesis: the production of cancerous cells.

cerebellum: the more primitive hindbrain responsible for basic functions of muscular control, breathing, circulation and balance.

cerebrum: the more recent front part of the brain, divided into two symmetrical hemispheres. In humans, it is the location of the higher functions, including reasoning, learning, sensory perception, and emotional response.

cetacea: a large sea mammal (e.g. a whale or a dolphin) that has a streamlined body with forelimbs modified as flippers, no hindlimbs and a blowhole on the back.

cognitive revolution: a period of time, starting around 70,000 years ago, associated with rapid increase in intellectual, social and artistic skills in *Homo sapiens.*

creationism: the belief that God created the universe.

Cro-Magnon: the first *Homo sapiens* to appear in Europe.

Crural Index: the ratio between the length of the shin bone and the thigh bone, used to indicate the overall shape of the body, whether slim or stout stature.

CT: computerized tomography imaging.

Denisovans: an extinct early hominid existing from 125,000–40,000 years ago for which remains were found in Siberia alongside archaeological remnants from *H. heidelbergensis* and Neanderthals.

DHA: docosahexaeonic acid, a lipoprotein, mainly found in the marine food chain, which has been an essential component vital for neural development for several hundred million years.

DNA: deoxyribonucleic acid. A nucleic acid molecule in the form of a twisted double strand double helix that is the major component of chromosomes and carries genetic information.

dominant: a term for describing a gene which, if present in a pair of genes, will dictate the characteristic in that organism and override any recessive gene.

endogenous: originating or growing within an organism or tissue.

eccrine: describes sweat glands that are distributed all over the body in humans, especially on the hands and feet, that do not secrete organic matter, and that are important in regulating body temperature.

ecological: environmental.

epigenetic: hereditary factors not controlled by genes.

epithelium: the surface lining of the skin and internal cavities such as the mouth and throat.

epoch: an era of time.

ethmoid sinus: air-filled cavities between the nose and orbits.

ethmo-turbinals: mucosa-covered folds (similar to maxillo-turbinals) inside the nose to increase the internal surface area to help with smell and humidification of inspired air.

exogenous: originating outside an organism or system.

exostoses: see surfer's ear.

gene: the basic unit capable of transmitting characteristics from one generation to the next. It consists of a specific sequence of DNA or RNA that occupies a fixed position locus on a chromosome.

genome: an organism's complete set of DNA present in all cells with a nucleus.

genotype: the genetic makeup of an organism, as opposed to its physical characteristics phenotype (appearance).

Great Rift Valley: a 40–60 km wide valley that stretches from the Afar peninsula, in northeast Africa near the Red Sea, to Mozambique. It formed around 6–7 million years ago when the east African tectonic plate split from the main continent, forming a vast inland sea. This is the site of many of the early hominid finds.

Hadar: one of the early hominid sites in the Great Rift Valley.

haemoglobin: an iron-containing protein in red blood cells that transports oxygen around the body.

haplogroup: a unique set of genes that characterise various species.

heterozygous: describes a cell or organism that has two or more different versions of at least one of its genes (e.g. one dominant and one recessive). The appearance of the offspring of such an organism will be determined by the dominant gene.

Holocene: the present era of time which started 10,000 years ago.

hominid: a primate belonging to a family that split from chimpanzees 6–7 million years ago, of which the modern human being is the only species still in existence.

hominin: a bipedal member of the hominid family with features approximating those of modern humans.

Homo erectus: an extinct ancestor of *H. sapiens* living approximately 1.8–30,000 years ago and known by the fossil record, initially found in Java, and thought to have had an upright stature, a moderate-sized brain and a low forehead.

Homo ergaster: an extinct ancestor of the *H. sapiens* living approximately 1.9–1.5 million years ago, with a stature similar to modern humans.

Homo floresiensis: an extinct ancestor of *H. sapiens* living approximately 95,000 to 12,000 years ago. Known as the 'hobbit' because of its short stature, fossils were found on the island of Flores in Indonesia. It was the last of the *Homo* family to become extinct, leaving only *H. sapiens*.

***Homo habilis*:** an extinct ancestor of *H. sapiens* living approximately 2.4–1.6 million years ago and characterised by the ability to make and use tools. Sometimes called 'handyman'.

***Homo heidelbergensis*:** an extinct ancestor of *H. sapiens* living approximately 600,000–200,000 years ago. Characterised by a large brain and a strong physique.

***Homo neanderthalensis*:** an extinct ancestor of *H. sapiens* living approximately 350,000–28,000 years ago. It was the first hominin for which fossils were discovered and mainly inhabited Europe.

***Homo sapiens*:** modern human beings that first appeared in Africa around 200,000 years ago and initially migrated into Europe and Asia. The last remaining extant hominin species whose populations have spread to every corner of the world – and beyond.

homozygous: describes a cell or organism that has identical versions of at least one of its genes (e.g. both dominant or both recessive). The appearance of the offspring of such an organism will be determined by the gene.

hybridization: gene combination (via either human intervention or naturally) of two different species or subspecies to generate a unique organism.

hypoxia: an inadequate supply of oxygen reaching the body's tissues.

Ice Age: a period in the Earth's history when temperatures fell worldwide and large areas of the Earth's surface were covered with glaciers. The most recent Ice Age occurred during the Pleistocene era (40,000–12,000 years ago) during which most of the northern hemisphere was covered with glaciers.

immunology: the scientific study of the way the immune system works in the body, including allergies, resistance to disease and acceptance or rejection of foreign tissue.

inter-canthal distance: the distance between two eye sockets.

introgression: the incorporation of genes from one species into the gene pool of another, resulting from hybridization and leading to the fusion of DNA from two separate genetic groups.

ischaemia: an inadequate supply of blood to a part of the body, caused by partial or total blockage or narrowing of an artery.

isotope: each of two or more forms of a chemical element with the same atomic number but different numbers of neutrons.

Koobi fora: one of the early hominid sites in the Great Rift Valley.

labyrinth: a structure consisting of connected cavities or canals in the ethmoid sinuses or inner ear.

miocene: the era of geological time, 24 million–5 million years ago, during which the modern ocean currents were established and Antarctica became frozen.

mitochondria: a small round or rod-shaped body that is found in the cytoplasm of most cells and produces enzymes for the metabolic conversion of food into energy.

morphological: the biology of the form and structure of an organism or of a part of an organism.

mousterian technology: developed during the 'Cognitive Revolution' by *H. neanderthalensis*, less sophisticated than the Aurignacian technology of *H. sapiens*.

mutation: a random change in a gene or chromosome resulting in a new trait or characteristic that can be inherited. Mutation can be a source of beneficial genetic variation, or can be neutral or harmful in effect.

myoglobin: an iron-containing protein resembling haemoglobin, found in muscle cells, that takes oxygen from the blood, releasing it to the muscles during strenuous exercise. The three-dimensional structure of myoglobin and the alpha and beta chains of haemoglobin are almost identical.

neanderthal: see *Homo neanderthalensis*.

Olduvai Gorge: one of the early hominid sites in the Great Rift Valley.

olfaction: the sense of smell.

Orrorin tugenensis: a contender for the earliest bipedal hominin, inhabiting lakeside areas in the Great Rift Valley, having features of bipedal locomotion, but also probably a tree climber.

osteum (a): small communication holes from the sinuses into the nasal cavity to allow air circulation and drainage.

pachyosteosclerosis: the condition of dense and brittle bones, characteristically seen in slow-moving semi-aquatic mammals.

palaeoanthropology: the study of early human beings and related species through fossil evidence.

papyrus: a tall water plant. It's flowers are small and umbrella-shaped. Native to southern Europe and the Nile valley.

phenotype: the visible characteristics of an organism resulting from the interaction between its genetic makeup and the environment.

physiological: relating to the way that living things function, rather than to their shape or structure.

pleistocene: the era of geological time, about 2.6 million to 11,000 years ago, characterised by the disappearance of continental ice sheets and the appearance of humans.

pliocene: the era of geological time, 5–2.6 million years ago, during which *H. erectus* first appeared.

pollination: the transfer of pollen grains from the male structure of a plant (anther) to the female structure of a plant (stigma) and subsequent fertilisation.

pre-hominid: members of the primate ape family in the lineage leading to hominids, but still mainly quadruped.

proboscis monkey: a large monkey with reddish fur and a protruding bulbous nose that in older males becomes pendulous. Native to Borneo.

pseudoscience: a theory or method doubtfully or mistakenly held to be scientific.

Ramapithecus: a pre-hominid ape.

recessive: genes carrying a certain characteristic that will only be expressed if there are no competing dominant genes for that characteristic.

Savannah theory: the traditional theory of human evolution that suggests early humans evolved on the savannah and stood upright because they could see further over the grass.

sedges: a wetland plant that resembles grass and has a triangular stem, leaves growing in three vertical rows and inconspicuous spikes of flowers.

Sehalanthropus tchadensis: an extinct ancestor of *H. sapiens* living approximately 6–7 million years ago, probably one of the last common ancestors for both humans and chimpanzees.

sinuses: otherwise known as paranasal sinuses, these are cavities filled with air in the bones of the face and skull, with a smallcommunicating opening into the nasal passages.

squalene: a hydrocarbon found in the liver that is an intermediate in the formation of cholesterol. It is also found in the skin of aquatic and semiaquatic mammals, including humans.

tectonic: relating to the forces that produce movement and deformation of the Earth's crust.

ungulate: a mammal with hoofs, e.g. the horse, rhinoceros, pig, giraffe, deer and camel.

vasoconstriction: narrowing of the blood vessels with consequent reduction in blood flow or increased blood pressure.

vernix caseosa: a white, greasy substance that covers the skin of a newborn human baby. It provides a protective waterproof coating in the womb. The only other mammal known to be born with a covering of vernix is the seal.

References

INTRODUCTION

---

1. Morgan, E. (1990). *The Scars of Evolution*. Souvenir Press, London, UK.

CHAPTER 1 THEORIES OF HUMAN EVOLUTION

---

1. Darwin, C. (1859). *The Origin of Species*. John Murray, London, UK.
2. Darwin, C. (1871). *The Descent of Man*. John Murray, London, UK.
3. Morgan, E. (1997). *The Aquatic Ape Hypothesis – The Most Credible Theory of Human Evolution*. Souvenir Press, London, UK.
4. Hardy, A.C. (1960). Was man more aquatic in the past? *The New Scientist*, 7: 642–645.
5. Tobias, P.V. (2011). Chapter 1. Revisiting water and hominid evolution. In: Vaneechoutte, M., Kuliukas, A. & Verhaegan, M. (Eds.). *Was Man More Aquatic in the Past? Fifty Years after Alister Hardy*. Bentham Books, Danvers, MA.
6. Attenborough, D. (2016). *The Waterside Ape*. BBC Publications. Radio 4: 14–15 September.
7. Rhys-Evans, P. (1992). The para-nasal sinuses and other enigmas: An aquatic evolutionary theory. *Journal of Laryngology & Otology*, 106: 214–225.
8. Rhys-Evans, P. & Cameron, M. (2014). Surfer's ear (aural exostoses) provides hard evidence of Man's aquatic past. *Human Evolution*, 29(1–3): 75–90.
9. Rhys-Evans, P. & Cameron, M. (2017). Aural exostoses provide vital fossil evidence of an aquatic phase in Man's early evolution. *Annals of the Royal College of Surgeons*, 99: 594–601.
10. Verhaegen, M., Munro, S., Puech, P.-F. & Vaneechoutte, M. (2011). Early hominoids: Orthograde aquarboreals in flooded forests? In: Vaneechoutte, M., Kuliukas, A. & Verhaegen, M. (Eds.). *Was Man More Aquatic in the Past? Fifty Years after Alister Hardy*. Bentham Science Publications, pp. 67–81.
11. Dart, R. (1925). *Australopithecus africanus*: The Man-Ape of South Africa. *Nature*, 115: 195–199.
12. Westenhofer, M. (1942). *Der Eigenweg des Menschen* (The Unique Road to Man).

13. Vrba, E.S. (1985). Environment and evolution: Alternative causes of the temporal distribution of evolutionary events. *South African Journal of Science*, 81: 229–236.
14. Potts, R. (1996). Evolution and climatic variability. *Science*, 273: 922–923.
15. Potts, R. (2013). Hominin evolution in settings of strong environmental variability. *Quaternary Science Reviews*, 73: 1–13.
16. Maslin, M.A., Brierley, C.M., Milner, A.M. et al. (2014). East African climate pulses and early human evolution. *Quaternary Science Reviews*, 101: 1–17.
17. Maslin, M.A. & Trauth, M.H. (2009). Plio-pleistocene East African pulsed climate variability and its influence on early human evolution. In: Grine, F.E., Leakey, R.E. & Fleagle, J.G. (Eds.). *The First Humans – Origins of the Genus Homo*. Springer Science, Berlin, Germany, pp. 151–158.
18. Schultz, S. & Maslin, M.A. (2013). Early human speciation, brain expansion and dispersal influenced by African climate pulses. *PLoS ONE*, 8(10): e76750.
19. Roberts, A. (2011). *Evolution – The Human Story*. Dorling Kindersley Ltd., London, UK.
20. Stringer, C. & Andrews, P. (2005). *The Complete World of Human Evolution*. Thames & Hudson, Ltd., London, UK.
21. Odent, M. (2017). *The Birth of Homo, the Marine Chimpanzee*. Pinter & Martin Ltd., London, UK.

CHAPTER 2 THE AQUATIC DEBATE

1. Morris, D. (1967). *The Naked Ape*. Jonathan Cape, London, UK.
2. Morgan, E. (1972). *The Descent of Women*. Souvenir Press, London, UK.
3. Morgan, E. (1985). *The Aquatic Ape Hypothesis*. Souvenir Press, London, UK.
4. Rhys-Evans, P. (1992). The para-nasal sinuses and other enigmas: An aquatic evolutionary theory. *Journal of Laryngology & Otology*, 106: 214–225.
5. Langdon, J.H. (1997). Umbrella hypotheses and parsimony in human evolution: A critique of the Aquatic Ape Hypothesis. *Journal of Human Evolution*, 33(4): 479–494.
6. Tobias, P.V. (2011). Chapter 1. Revisiting water and hominid evolution. In: Vaneechoutte, M., Kuliukas, A. & Verhaegan, M. (Eds.). *Was Man More Aquatic in the Past? Fifty Years after Alister Hardy*. Bentham Books.
7. Gee, H. (2013). Aquatic apes are the stuff of creationism, not evolution. *Occam's Corner, The Guardian*, 9 May.
8. Hawks, J.D. (2009). Why anthropologists don't accept the Aquatic Ape Theory. *Blog post*, 4 August.
9. Frazier, K. (2015). Quacks and Cranks; GMO (genetically modified organisms) and climate, science and philosophy – CFI conference covers it all. *Skeptical Inquirer*, 39(5): 12.
10. Foley, R. & Lahr, M.M. (2014). The role of 'the aquatic' in human evolution – constraining the Aquatic Ape Hypothesis. *Evolutionary Anthropology*, 23: 56–59.

11. Attenborough, D. (2016). *The Waterside Ape*. BBC Publications. Radio 4: 14–15 September.
12. Roberts, A. & Maslin, M.A. (2016). Sorry David Attenborough, we didn't evolve from 'aquatic apes' – Here's why. *TheConversation.com*.
13. Schagatay, E., Rhys-Evans, P., Stewart, K. et al. (2016). A reply to Alice Roberts and Mark Maslin: Our ancestors may indeed have evolved at the shoreline – and here is why. Technical Report. doi:10.13140/RG.2.2.23127.68007.
14. Johanson, D.C. & Shreeve, J. (1989). *Lucy's Child: The Discovery of a Human Ancestor*. Early Man Publishing, New York.
15. Sarich, V.M. & Wilson, A.C. (1967). Immunological time scale for hominid evolution. *Science*, 158: 1200–1203.
16. Johanson, D.C. & White, T.D. (1979). A systematic assessment of early African Hominids. *Science*, 203: 321–330.
17. Bonatti, E., Emiliani, C., Ostlund, G. & Rydell, H. (1971). Final dessication of the Afar Rift, Ethiopia. *Science*, 172: 468–469.
18. Tazieff, H. (1972). Tectonics of central Afar. *Journal of Earth Science*, 8(2): 171–182.
19. Johanson, D. & Edey, M. (1981). *Lucy – The Beginnings of Humankind*. Simon & Schuster, New York.
20. Rightmire, G.P. (1990). *The Evolution of Homo Erectus*. Cambridge University Press, Cambridge, UK.
21. Morgan, E. (1990). *The Scars of Evolution*. Souvenir Press, London, UK.
22. Rhys-Evans, P. & Cameron, M. (2014). Surfer's ear (aural exostoses) provides hard evidence of Man's aquatic past. *Human Evolution*, 29(1–3): 75–90.
23. Rhys-Evans, P. & Cameron, M. (2017). Aural exostoses provide vital fossil evidence of an aquatic phase in Man's early evolution. *Annals of the Royal College of Surgeons*, 99(8): 594–601.
24. Gee, H. (2016). *Why On Earth Is the BBC Stooping Once Again?* Twitter Account.

CHAPTER 3 OUR GENETIC HERITAGE

1. Jones, S. (2009). *Darwin's Island*. Little, Brown Book Group, London, UK.
2. Darwin, C. (1871). *The Descent of Man*. John Murray, London, UK.
3. Mendel, G. (1866). Versuche über Pflanzenhybriden. In: *Verhandlungen des naturforschenden Vereins Brünn*. Im Verlage des Vereines, Brunn, Germany.
4. Watson, J.D. & Crick, F.H.C. (1953). A structure for deoxyribose nucleic acid. *Nature*, 171: 737–738.
5. Dupont, L.M., Rommerskirchen, F., Mollenhauer, G. & Schefuß, E. (2013). Miocene to Pliocene changes in South African hydrology and vegetation in relation to the expansion of C_4 plants. *Earth and Planetary Science Letters*, 375(1): 408–417.
6. Tolley, K.A., Chase, B.M. & Forest, F. (2008). Speciation and radiations track climate transitions since the Miocene Climatic Optimum: A case study of southern African chameleons. *Journal of Biogeography*, 35: 1402–1414.
7. Leakey, R.E. (1981). *The Making of Mankind*. E.P. Dutton, New York.

CHAPTER 4 OUR EARLY ANCESTORS

1. Roberts, A. (2011). *Evolution – The Human Story*. Darling Kindersley, London, UK.
2. Johanson, D. & Shreeve, J. (1989). *Lucy's Child: The Discovery of a Human Ancestor*. Early Man Publishing, New York.
3. Lewin, R. (2005). *Human Evolution – An Illustrated Introduction*. Blackwell Publishing, Oxford, UK.
4. McKie, R. (1999). *Ape Man – The Story of Human Evolution*. BBC Publications.
5. Morgan, E. (1990). *The Scars of Evolution*. Souvenir Press, London, UK.
6. Rightmire, G. (1998). Human evolution in the Middle Pleistocene: The role of *Homo heidelbergensis*. *Evolutionary Anthropology: Issues, News, and Reviews*, 6: 218–227.
7. Mounier, A., Marchal, F. & Condemi, S. (2009). Is *Homo heidelbergensis* a distinct species? New insight on the Mauer mandible. *Journal of Human Evolution*, 56(3): 219–246.
8. Meredith, M. (2011). *Born in Africa: The Quest for the Origins of Human Life*. Public Affairs, New York.
9. Rogers, A.R., Bohlender, R.J. & Huff, C.D. (2017). Early history of Neanderthals and Denisovans. *Proceedings of the National Academy of Sciences*, 114(37): 9859–9863.

CHAPTER 5 THE NEANDERTHALS AND THEIR DEMISE

1. Rogers, A.R., Bohlender, R.J. & Huff, C.D. (2017). Early history of Neanderthals and Denisovans. *Proceedings of the National Academy of Sciences*, 114(37): 9859–9863.
2. Wallace, A.R. (1864). The origin of human races and the antiquity of man deduced from the theory of natural selection. *Journal of the Anthropological Society of London*, 2(CLXIV): clviii–clxxxvii.
3. Walker, A. & Shipman, P. (1996). *The Wisdom of the Bones: In Search of Human Origins*. Weidenfield & Nicolson, London, UK.
4. Stewart, J.R. & Stringer, C.B. (2012) Human evolution out of Africa: The role of refugia and climate change. *Science*, 335(6074): 1317–1321. doi:10.1126/science.1215627.
5. Mounier, A., Marchal, F. & Condemi, S. (2009). Is *Homo heidelbergensis* a distinct species? New insight on the Mauer mandible. *Journal of Human Evolution*, 56(3): 219–246.
6. Rutherford, A. (2016). *A Brief History of Everyone Who Ever Lived*. Weidenfeld & Nicolson, London, UK.

7. McKie, R. (1999). *Ape Man – The Story of Human Evolution*. BBC Publications.
8. Hoffman, D.L., Standish, C.D., Garcia-Diez, M., Pettitt, P.B., Milton, J.A. & Zilhao, J. (2018). U-Th dating of carbonate crusts reveals Neandertal origin of Iberian cave art. *Science*, 359(6378): 912–915.

CHAPTER 6 THE WATERSIDE APE: WHY ARE WE SO DIFFERENT?

1. Bronowski, J. (1974). *The Ascent of Man*. BBC, London, UK, p. 26.
2. Hardy, A.C. (1960). Was Man more aquatic in the past? *New Scientist*, 7: 642–645.
3. Roede, M., Wind, J., Patrick, J. and Reynolds, V. (1991). *The Aquatic Ape: Fact or Fiction?* Souvenir Press, London, UK.
4. Darwin, C. (1871). *The Descent of Man*. John Murray, London, UK.
5. Sokalov, V.E. (1983). *Mammal Skin*. University of California, Berkeley, CA.
6. Morgan, E. (1990). *The Scars of Evolution*. Souvenir Press, London, UK.
7. Pond, C. (1997). The biological origin of adipose tissue in humans. In: Morbeck, M.E., Galloway, A. & Zihlman, A. (Eds.). *The Evolving Female*. Princeton University Press, Princeton, NJ.
8. Lewin, R. (2005). *Human Evolution – An Illustrated Introduction*. Blackwell Publishing, London, UK.
9. Ruff, C.B. (1994). Morphological adaptation to climate in modern and fossil hominids. *American Journal of Physical Anthropology*, 95(Suppl. 19): 65–107.
10. Montagna., W. and Parakkal, P.F. (1974). *The Structure and Function of Skin*. Academic Press Inc., New York.
11. Roebroeks, W. (2007). *An Integrative Approach to the Hominin Record*. Leiden University Press, Leiden.

CHAPTER 7 THE NAKED APE

1. Changizi, M., Weber, R., Kotecha, R. & Palazzo, J. (2011). Are wet-induced wrinkled fingers primate rain treads? *Brain, Behaviour and Evolution*, 77: 286–290.
2. Kareklas, K., Nettle, D. & Smulders, T.V. (2012). *Biology Letters*, 10, 1098.
3. Darwin, C. (1871). *Descent of Man*. John Murray, London, UK.
4. Sokalov, V. (1983). *Mammal Skin*. University of California, Berkeley, CA.
5. Morgan, E. (1990). *The Scars of Evolution*. Souvenir Press, London, UK.

CHAPTER 8   WHY WE LOST OUR COATS: THE EARLY HOMININ TAILORS

1. Morris, D. (1967). *The Naked Ape.* Jonathan Cape, London, UK.
2. Rhys-Evans, P. & Cameron, M. (2013). Aural exostoses provide vital evidence of an aquatic phase in Man's early evolution. *Human Evolution,* 29(1–3): 75–90.
3. Sokalov, V. (1983). *Mammal Skin.* University of California, Berkeley, CA.
4. Precht, H., Christophersen, J., Hensel, H. & Larcher, W. (Eds.). (1973). *Temperature and Life.* Springer, Berlin, Germany.
5. Hiley, P.G. (1976). The thermoregulatory responses of the Galago, the Baboon and the Chimpanzee to heat stress. *Journal of Physiology,* 254: 657–671.
6. Maloiy, G.M.O., Rugangazi, B.M. & Clemens, E.T. (1988). Physiology of the Dik-Dik antelope. *Comparative Biochemistry and Physiology,* 91(1): 1–8.
7. Wheeler, P. (1992). The influence of the loss of functional body hair on the water budgets of early hominids. *Journal of Human Evolution,* 23(5): 379–388.
8. Ruxton, G.D. and Wilkinson, D.M. (2011a). Avoidance of overheating and selection for both hair loss and bipedality in hominins. *Proceedings of the National Academy of Sciences USA,* 108: 20965–20969.
9. Ruxton, G.D. and Wilkinson, D.M. (2011b). Thermoregulation and endurance running in extinct hominins: Wheeler's models revisited. *Journal of Human Evolution,* 61: 169–175.
10. Morgan, E. (1972). *The Descent of Woman.* Souvenir Press, London, UK.
11. Kittler, R., Kayser, M. & Stoneking, M. (2003). Molecular evolution of Pediculus humanis and the origin of clothing. *Current Biology,* 13(16): 1414–1417.
12. Klein, R.G. (2002). *The Dawn of Human Culture.* John Wiley & Sons, London, UK.
13. Rogers, A. (2017). Early history of Neanderthals and Denisovans. *Proceedings of the National Academy of Science USA.* doi:10.1073/pnas.1706426114.
14. Harding, R. (2000). Evidence of variable selective pressures at MC1R. *American Journal of Human Genetics,* 66(4): 1351–1361.
15. Jablonski, N.G. & Chaplin, G. (2010). Human skin pigmentation as an adaptation to UV radiation. *Proceedings of the National Academy of Sciences USA,* 107: 8962–8968.
16. Kostyuk, V., Potapovich, A., Stancato, A. et al. (2012). Photo-oxidation products of skin surface squalene mediate metabolic and inflammatory responses to solar UV in human keratinocytes. *PLoS One,* 7(8): e44472.
17. Uribe, P. & Gonzalez, S. (2011). Epidermal Growth Factor Receptor (EGFR) and squamous cell carcinoma of the skin: Molecular bases for EGFR-targeted therapy. *Pathology Research and Practice,* 207(6): 337–342.
18. O'Charoenrat, P., Rhys-Evans, P., Modjtadedi, H., Court, W., Box, G. & Eccles, S.A. (2000). Over expression of EGFR in human HNSCC cell lines correlates with matrix metalloproteinase-9 expression in *in-vitro* invasion. *International Journal of Cancer,* 86(3): 307–317.
19. O'Charoenrat, P., Rhys-Evans, P., Modjtahedi, H., Box, G.M. & Eccles, S.A. (2000). Epidermal growth factor like ligands differentially upregulate matrix metalloproteinase-9 in HNSCC cells. *Cancer Research,* 60: 1121–1128.

20. Khode, S.R., Rhys-Evans, P., Dwivedi, R.C. & Kazi, R. (2014). Exploring the link between HPV and oral and oro-pharyngeal cancers. *Journal of Cancer Research and Therapeutics*, 10(3): 492–498.

21. Gaffney, D.C., Soyer, H.P. & Simpson, F. (2014). The EGFR in squamous cell carcinoma: An emerging drug target. *Australasia Journal of Dermatology*, 55(1): 24–34.

22. Rogers, S.J., Rhys-Evans, P., Box, C. et al. (2009). Determinants of response to EGFR tyrosine kinase in HNSCC. *Journal of Pathology*, 218(1): 122–130.

23. Dwivedi, R.C., Rhys-Evans, P., Kazi, R., Kanwar, N., Nutting, C.M. & Harrington, K.J. (2011). Should the treatment paradigms for oral and oropharyngeal cancers be changed now? The role of human papilloma virus. *ANZ Journal of Surgery*, 81(9): 581–583.

24. Greaves, M. (2014). Was skin cancer a selective force for black pigmentation in early hominin evolution? *Proceedings of the Royal Society B*, 281: 2132955.

CHAPTER 9 EVOLUTIONARY ADAPTATIONS IN THE HUMAN SKULL AND SINUSES

1. Morgan, E. (1990). *The Scars of Evolution*. Souvenir Press, London, UK.

2. Versalius, A. (1543). *De Humani Corporis Fabrica*. Basel, Switzerland.

3. Negus, V. (1958). *Comparative Anatomy and Physiology of the Nose and Paranasal Sinuses*. Livingstone, London, UK.

4. Blaney, S.P.A. (1990). Why paranasal sinuses? *Journal of Laryngology and Otology*, 104: 690–693.

5. Takahashi, R. (1984). The formation of human paranasal sinuses. *Acta Otolaryngologica*, 97(Supplement 408): 1–28.

6. Hardy, A. (1960). Was man more aquatic in the past? *The New Scientist*, 7: 642–645.

7. Rhys-Evans, P. (1992). The paranasal sinuses and other enigmas: An aquatic evolutionary theory. *Journal of Laryngology and Otology*, 106: 214–225.

8. Wood-Jones, F. (1916). *Arboreal Man*. E. Arnold, London, UK.

9. Cave, A.J.E. & Haines, R.W. (1940). Paranasal sinuses of the anthropoid apes. *Journal of Anatomy*, 74: 493–523.

10. Mosher, H.P. (1929). Symposium on the ethmoid: The surgical anatomy of the ethmoidal labyrinth. *Transactions of the American Academy of Ophthalmology*, 34: 376–410.

11. Bartholinus, T. (1660). *Anatomica ex Caspari Bartholini, parentis institutionibus, omnique recentiorum et propriis observationibus, tertium ad sanguinis circulationem reformata cum iconibus novis accuratissimis*. Hagae Comitis, p. 488.

12. Proetz, A.W. (1953). *Applied Physiology of the Nose*, 2nd ed. Annals Publishing Company, St. Louis, MO.

13. Haller, A. (1763). *Elementa physiologicae corporis humani.* Liber XIV, 5: p. 180. (Cited by Wright, p. 169, 1914).
14. Skillern, R.H. (1920). *The Accessory Sinuses of the Nose,* 2nd ed. Lippencott, Philadelphia, PA.
15. Mygind, N. & Winther, B. (1987). Immunological barriers in the nose and paranasal sinuses. *Acta Otolaryngologica* 103: 363–368.
16. Cloquet, H. (1830). *A System of Human Anatomy* (translation by Robert Knox, 1838). Maclachlan & Stewart, Edinburgh, UK, p. 582.
17. Rhys-Evans, P. (1987). Anatomy of then nose and paranasal sinuses. In: Wright, D.A. & Kerr, A.G. (Eds.). *Scott-Brown's Otolaryngology, Volume 1:Basic Science,* 5th ed. Butterworths, London, UK.
18. Brothwell, D.R., Molleson, T. & Metreweli, C. (1968). Radiological aspects of normal variation in early skeletons: An exploratory study. The skeletal biology of early human populations. In: *Society for the Study of Human Biology,* Vol. 8. Pergammon Press, Oxford, UK.
19. Wolfowitz, B.L. (1974). *Pneumatization of the Skull of the Southern African Negro.* PhD thesis, University of Witwaterstrand.
20. Kowertvelyessy, T. (1972). Relationship between the frontal sinus and climatic conditions. A skeletal approach to cold adaptations. *American Journal of Physical Anthropology,* 37: 161–173.
21. Tillier, A.M. (1975). *Les sinus crannies chez les hommes actuels et fossils: essai d'interpretation.* PhD thesis, University of Paris.
22. Coon, C.S. (1962). *The Origin of Races.* Knopf, New York.
23. Proetz, A.W. (1922). Observations upon the formation and function of the accessory nasal sinuses and mastoid cells. *Annals of Otology, Rhinology and Laryngology,* 39: 1083–1100.
24. Shea, B.T. (1985). On aspects of skull form in African apes and orangutans with implications for hominid evolution. *American Journal of Physical Anthropology,* 68: 329–342.
25. Morgan, E. (1972). *The Descent of Woman.* Souvenir Press, London, UK.
26. Morgan, E. (1982). *The Aquatic Ape.* Souvenir Press, London, UK.

CHAPTER 10 HUMAN SKULL BUOYANCY AND THE DIVING REFLEX

1. Skillern, R.H. (1920). *The Accessory Sinuses of the Nose,* 2nd ed. Lippencott, Philadelphia, PA.
2. Negus, V. (1958). *Comparative Anatomy and Physiology of the Nose and Paranasal Sinuses.* Livingstone, London, UK.
3. Rhys-Evans, P. (1992). The paranasal sinuses and other enigmas: An aquatic evolutionary theory. *Journal of Laryngology and Otology,* 106: 214–225.
4. Lindholm, P. & Lundgren, C.E.G. (2009). The physiology and pathophysiology of human breath-hold diving. *Journal of Applied Physiology,* 106(1): 284–292.

5. Rybka, E.J. & McCulloch, P.F. (2006). The anterior ethmoidal nerve is necessary for the initiation of the nasopharyngeal response in the rat. *Brain Research*, 1075: 122–132.

6. Seymour R.S., Bosiocic, V. & Snelling E.P. (2016). Fossil skulls reveal that blood flow rate to the brain increased faster than brain volume during human evolution. *Royal Society Open Science*, 3: 160305.

7. Kooyman, G.L. (1989). *Diverse Divers: Physiology and Behaviour.* Springer-Verlag, Berlin, Germany.

8. Noren, S.R., Williams, T.N., Pabst, D.A. et al. (2001). The development of diving in marine endotherms: Preparing the skeletal muscles of dolphins, penguins and seals for activity during submergence. *Journal of Comparative Physiology B*, 171: 127–134.

9. Llardo, M.D., Moltke, I., Korneliussen, T.S. et al. (2018). Physiological and genetic adaptations to diving in Sea Nomads. *Cell*, 173(3): 569–580.

10. Lundberg, J.O., Farkas-Szallasi, T., Weitzberg, E. et al. (1995). High nitric oxide production in human paranasal sinuses. *Nature Medicine*, 1: 370–373.

11. Weitzberg, E. & Lundberg, J.O. (2002). Humming greatly increases nasal nitric oxide. *American Journal of Respiratory and Critical Care Medicine*, 166: 144–145.

12. Cardell, L.O. (2002). The paranasal sinuses and a unique role in airway nitric oxide production? *American Journal of Respiratory and Critical Care Medicine*, 166(2): 131–132.

13. Lapi, D., Scuri, R. & Colantuoni, A. (2016). Trigeminal cardiac reflex and cerebral blood flow regulation. *Frontiers in Neuroscience*, 10: 470.

14. Young, J.Z. (1950). *The Life of Vertebrates.* Clarendon Press, Oxford, UK.

15. Thompson, A. & Dudley-Buxton, L.H. (1923). Man's nasal index in relation to certain climatic conditions. *Journal of the Royal Anthropological Institute*, 53: 92–122.

CHAPTER 11   SURFER'S EAR

1. Hardy, A.C. (1960). Was man more aquatic in the past? *New Scientist*, 7: 642–645.

2. Morgan, E. (1990). *The Scars of Evolution.* Souvenir Press, London, UK.

3. Morgan, E. (1997). *The Aquatic Ape Hypothesis.* Souvenir Press, London, UK.

4. Rhys-Evans, P. (1992). The paranasal sinuses and other enigmas: An aquatic evolutionary theory. *The Journal of Laryngology and Otology*, 106: 214–225.

5. Rhys-Evans, P. & Cameron, M. (2014). Surfer's ear (aural exostoses) provides hard evidence of Man's aquatic past. *Human Evolution*, 29: 75–90.

6. Rhys-Evans, P. & Cameron, M. (2017). Aural exostoses provide vital fossil evidence of an aquatic phase in early human evolution. *Annals of the Royal College of Surgeons of England*, 99(8): 594–601.

7. Roche, A.F. (1964). Aural exostoses in Australian Aboriginal skulls. *Annals of Otology, Rhinology and Laryngology*, 73: 82–91.

8. Herdlycke, A. (1935). Ear exostoses. *Smithsonian Miscellaneous Collection*, 93: 1–100.

9. Gregg, J.B. and Bass, W.H. (1970). Exostoses in the external auditory canals. *Annuals of Otology, Rhinology and Layyngology*, 79: 834–839.

10. Gregg, J.B. and Gregg, P.S. (1988). *Dry Bones. Dakota Territory Reflected.* University of South Dakota Press, Vermillion, SD.

11. Aufderheide, A.C., Rodriguez, C. & Langsjoen, O. (1998). *The Cambridge Encyclopaedia of Human Paleopathology.* Cambridge University Press, Cambridge, UK.

12. Okumura, M.M., Boyadjan, H.C. & Eggers, S. (2007). Auditory exostoses as an aquatic activity marker: A comparison of coastal and inland skeletal remains from tropic and subtropical regions of Brazil. *American Journal of Physical Anthropology*, 132: 558–567.

13. Van Gilse, P.H.G. (1938). Des observations ulterieurssur la genese des exostoses du conduit externe par l'iritation d'eau froid. *Acta Oto-laryngologica*, 26: 343–352.

14. Ascenzi, A. & Balistreri, P. (1975). Aural exostoses in a Roman skull excavated at the 'Baths of the Swimmer' in the ancient town of Ostia. *Journal of Human Evolution*, 4: 579–584.

15. Kennedy, G.E. (1986). The relationship between auditory exostoses and cold water: a latitudinal analysis. *American Journal of Physical Anthropology*, 71: 401–415.

16. Deleyiannis, F.W., Cockroft, B.D. & Pinczower, E.F. (1996). Exostoses of the external ear canal in Oregan surfer's. *American Journal of Otolaryngology*, 17: 303–307.

17. Hutchinson, D.L., Denise, C.B. et al. (1997). A re-evaluation of the cold water etiology of external auditory exostoses. *American Journal of Physical Anthropology*, 103: 417–422.

18. Ballachandra, B. (1994). *The Human Ear Canal.* Singular Publishing Group, San Diego, CA.

19. Kerr, A.G. (1997). *Scott Brown's Otolaryngology, Embryology,* 6th ed. CRC Press, Boca Raton, FL.

20. Michaels, L. (1997). The ear. In: Sternberg, E. (Ed.). *Histology for Pathologists*, 2nd ed. Lippincott-Raven, New York, pp. 337–366.

21. Pilch, B.Z. (2001). *Head and Neck Surgical Pathology.* Lippincott Williams & Wilkins, Philadelphia, PA, p. 59.

22. Fowler, E.P. & Osman, P.M. (1942). New bone growth due to cold water in the ears. *Archives of Otolaryngology*, 6: 455–456.

23. Harrison, D.F.N. (1962). The relationship of osteomata of the external auditory meatus to swimming. *Annals of the Royal College of Surgeons of England*, 31: 187–201.

24. van den Broek, A.J.P. (1943). On exostoses in the human skull. *Acta Neerland Morphol*, 5: 95–118.

25. Lerner, U.H. (1994). Regulation of bone metabolism by the Kalllicrein-kinin system, the coagulation cascade and the acute phase reactants. *Oral Surgery, Oral Medicine, Oral Pathology*, 78: 481–493.

26. Mundy, G.R., Boyce, B., Hughes, D. et al. 1995. The effects of cytokines and growth factors on osteoblastic cells. *Bone*, 17(2 Suppl): 71–75.

27. Luo, Z.X., Ruf, I. & Martin, T. (2012). The petrosal and inner ear of the late Jurassic cladotherian mammal *Dryolestes leiriensis* and implications for ear evolution in therian mammals. *Zoological Journal of the Linnean Society*, 166: 433–463.

28. Stenfors, L.A., Sade, J., Hellstrom, S. & Anniko, M. (2001). How can the hooded seal dive to a depth of 1000m without rupturing its tympanic membrane? *Acta Otolaryngologica*, 121: 689–695.

29. Belgraver, P. (1935). Over exostoses van de Uitwengige Gehoorgang. Luctor, Leiden, the Netherlands.

30. Field, G.P. (1893). *A Manual of Diseases of the Ear*. Balliere, Tendall, and Cox, London, UK.

31. Korner, O. (1904). Ober den angeblich zyklischen Verlauf der akuten Paukenhohlenentzundung. *Z. Ohrenheilkd*, 46: 369–372.

32. Wong, B.J.F., Cervantes, W., Doyle, K.J., Karamzadeh, A.M., Boys, P., Brauel, G. & Mushtaq, E. (1999). Prevalence of external auditory canal exostoses in surfers. *Archives of Otolaryngology – Head & Neck Surgery*, 125(9): 969–972.

33. Mann, G.E. (1986). The torus auditivus: A reappraisal. *Paleopathology Newsletter*, 53: 5–9.

34. Moore, R., Schuman, T.A., Scott, T.A., Mann, S.E., Davidson, M.A., Labadie, R.F. (2010). Exostoses of the external auditory canal in white water kayakers. *The Laryngoscope*, 120: 582–590.

35. Verhaegen, M. (1993). Aquatice versus Savanna: comparative and paleo-environmental evidence. *Nutrition and Health*, 9(3): 165–191.

36. Kroon, D.F., Lawson, M.L. et al. (2002). Surfer's ear: External auditory exostoses are more prevalent in cold water surfers. *Otolaryngology Head & Neck Surgery*, 126: 499–504.

37. Barr, T. (1901). *Manual of Diseases of the Ear: Including Those of the Nose and Throat in Relation to the Ear for the Use of Students and Practitioners of Medicine*. James Maclehose, Glasgow, UK.

38. Verhaegen, M. & Munro, S. (2011). Pachyosteosclerosis suggests archaic Homo frequently collected sessile littoral foods. *Journal of Comparative Human Biology*, 62: 237–247.

39. Perez, P.J., Gracia, A., Martinez, I. & Arsuaga, J.L. (1997). Paleopathological evidence of the cranial remains from the Sima de los Huesos Middle Pleistocene site (Sierra de Atapuerca, Spain). Description and preliminary inferences. *Journal of Human Evolution*, 33: 409–421.

40. Joordens, J.C.A., Wesselingh, F.P., de Vos, J., Vonhof, H.B. & Kroon, D. (2009). Relevance of aquatic environments for hominins: A case study from Trinil (Java, Indonesia). *Journal of Human Evolution*, 57: 656–671.

41. Braun, D.R., Harris, J.W.K., Levin, N.E. et al. (2010). Early hominin diet included diverse terrestrial and aquatic animals 1.95 Myr ago in East Turkana, Kenya. *Proceedings of the National Academy of Sciences USA*, 107: 10002–10007.

42. Rightmire, G.P. (1990). *The Evolution of Homo erectus*. Cambridge University Press, Cambridge, UK.

43. Verhaegen, M. (1991). Aquatic features in fossil hominids. In: Roede, M., Wind, J., Patrick, J. & Reynolds, V. (Eds.). *Aquatic Ape: Fact or Fiction?* Souvenir Press, London, UK, p. 83.

CHAPTER 12 EVOLUTION OF THE HUMAN BRAIN

1. Kaas, J.H. (2013). The evolution of brains from early mammals to humans. *Wiley Interdisciplinary Reviews: Cognitive Science*, 4(1): 33–45.
2. Hofman, M.A. (2001). Evolution and complexity of the human brain: Some organizing principles. In: Roth, G. & Wullimman, M.F. (Eds.). *Brain Evolution and Cognition*. Wiley, New York, pp. 501–552.
3. Bernd, F. (1998). Of mice and others: Evolution of vertebrate brain development. *Brain, Behavior and Evolution*, 52: 207–217.
4. Rosa-Molinar, E. & Pritz, M.B. (2005). *Hindbrain Evolution, Development and Organisation*. Karger, San Diego, CA.
5. Rilling, J.K. (2014). Comparative primate neuroimaging: Insights into human brain evolution. *Trends in Cognitive Sciences*, 18: 45–55.
6. Crawford, M.A., Hussein, I., Nyuar, K.B. & Broadhurst, C.L. (2014). The global crisis in brain nutritionand the rise in mental ill-health. *Human Evolution*, 29(1–3): 207–227.
7. Brenna, J.T., Salem, Jr. N., Sinclair, A.J. & Cunnane, S.C. (2009). Alpha-linoleic acid supplementation and conversion to n-3 long-chain polyunsaturated fatty acids in humans. *Postaglandins Leukot Essent Fatty Acids*, 80(2–3): 85–91.
8. Williams, G. & Crawford, M.A. (1987). Comparison of the fatty acid component in structural lipids from dolphins, zebra and giraffe: Possible evolutionary implications. *Journal of Zoology London*, 213: 673–684.
9. Broadhurst, C.L., Wang, Y., Crawford, M.A. et al. (2002). Brain-specific lipids from marine, lacustrine, or terrestrial food sources: Potential impact on early African Homo sapiens. *Comparative Biochemistry and Physiology Part B, Molecular Biology*, 131: 653–673.
10. Morwood, M.J., O'Sullivan, P.B., Aziz, F. & Raza, A. (1998). Fission track ages of the stone tools and fossils on the east Indionesian island of Flores. *Nature*, 392: 173–176.
11. Klein, R.G., Avery, G., Cruz-Uribe, K. et al. (2004). The Ysterfontein 1 Middle Stone Age site, South Africa, and early human exploitation of coastal resources. *Proceedings of the National Academy of Sciences USA*, 101(16): 5708–5715.
12. Marean, C.W., Bar-Matthews, M., Bernatchez, J. et al. (2007). Early human use of marine resources and pigment in South Africa during the Middle Pleistocene. *Nature*, 449(7164): 905–908.
13. Stringer, C. (2000). Coasting out of Africa. *Nature*, 405: 24–27.
14. Leigh, S.R. (2004). Brain growth, life history and cognition in primate and human evolution. *American Journal of Primatology*, 62: 139–164.
15. McGraw, M.B. (1939). *The Neuromuscular Maturation of the Human Infant*. Institute of Child Development, New York.

CHAPTER 13 FOOD FOR THOUGHT AND THE COGNITIVE REVOLUTION

1. Darwin, C. (1871). *The Descent of Man.* John Murray, London, UK.
2. Seymour, R.S., Bosiocic, V. & Snelling E.P. (2016). Fossil skulls reveal that blood flow rate to the brain increased faster than brain volume during human evolution. *Royal Society Open Science*, 3: 160305.
3. Gibbons, A. (2007). Food for thought: Did the first cooked meals help fuel the dramatic evolutionary expansion of the human brain? *Science*, 316(5831): 1558–1560.
4. Harari, Y.N. (2014). *Sapiens: A Brief History of Humankind.* Harvill Secker, London, UK.
5. Rampiro, M.R. & Self, S. (1992). Volcanic winter and accelerated glaciations following the Toba super-eruption. *Nature*, 359(6390): 50–52.
6. Smith, E.I., Jacobs, Z., Johnsen, R. et al. (2018). Humans thrived in South Africa through the Toba eruption about 74,000 years ago. *Nature*, 2018. doi:10.1038/nature25967.
7. Cross, I., Zubrow, E. & Cowan, F. (2002). Musical behaviours and the archaeological record: A preliminary study. In: Mathieu, J. (Ed.). *Experimental Archaeology.* British Archaeological Reports International Series, pp. 25–34.
8. Cook, J. (2017). The Lion Man: An Ice Age masterpiece. *Blog*, 10 October, The British Museum.
9. Dunbar, R.I.M. (1998). The social brain hypothesis. *Evolutionary Anthropology: Issues, News and Reviews*, 6(5): 178–190.
10. David-Barrett, T. & Dunbar, R.I.M. (2013). Processing power limits social group size: Computational evidence for the cognitive costs of sociality. *Proceedings of the Royal Society B*, 280(1765): 20131151.
11. Evans, P.D., Gilbert, S.L., Mekel-Bobroz, N. et al. (2005). Microcephalin, a gene regulating brain size, continues to evolve adaptively in humans. *Science*, 309(5741): 1717–1720.

CHAPTER 14 THE HUMAN LARYNX AND EVOLUTION OF VOICE

1. Laitman, J.T. & Reidenberg, J.S. (1993). *Comparative and Developmental Anatomy of Laryngeal Position*, Vol. 1. J.B. Lippincott Co., Philadelphia, PA.
2. Lieberman, P. (1984). *The Biology and Evolution of Language.* Harvard University Press, Cambridge, MA.

3. Keleman, G. (1963). Comparative anatomy and performance of the vocal organ in vertebrates. In: Busnel, R. (Ed.). *Acoustic Behaviour of Animals.* Elsevier, Amsterdam, the Netherlands, pp. 489–521.
4. Gould, S.J. & Vrba, E.S. (1982). Exaptation – A missing term in the science of form. *Paleobiology,* 8: 4–15.
5. Riede, T., Bronson, E., Hatzikirou, H. & Zuberbuhler, K. (2005). Vocal production mechanisms in a non-human primate: Morphological date and a model. *Journal of Human Evolution,* 48(1): 85–96.
6. Bergeson, P.S. & Shaw, J.C. (2001). Are infants really obligatory nasal breathers? *Clinical Paediatrics,* 40(10): 567–569.
7. Fitch, W.T. (2000). The evolution of speech: A comparative review. *Trends in Cognitive Science,* 4(7): 258–267.
8. Lieberman, P. (2007). The evolution of human speech: Its anatomical and neural bases. *Current Anthropology,* 48(1): 39–66.
9. Ohala, J.J. (2000). The irrelevance of the lowered larynx in modern Man for the development of speech. In: *The Evolution of Language.* ENST, Paris, France, pp. 171–172.
10. Fahlman, A. & Schagatay, E. (2014). Man's place among the diving mammals. *Human Evolution,* 29(1–3): 47–66.
11. Schagatay, E. & Lodin-Sundstrom, A. (2011). Underwater working times in two groups of traditional apnea divers in Asia: The Ama and the Bajau. *Diving and Hyperbaric Medicine,* 41(1): 27–30.
12. Geschwind, N. & Levitsky, W. (1968). Human brain left-right asymmetries in temporal speech region. *Science,* 161: 186–187.
13. Gannon, P.J., Holloway, R.L., Broadfield, D.C. & Braun, A.R. (1998). Asymmetry of chimpanzee Planum Temporale: Humanlike pattern of Wernicke's brain language area homolog. *Science,* 279: 220–222.
14. Spiers, H., Emo, B. & Javadi, A.-H. (2017). Satnavs 'switch off' parts of the brain. *Nature Communications,* 21 March.
15. Laitman, J.T. (1986). L'origine du langage articulé. *La recherche.,* 181, 17: 1165–1173 (Figs. 1 and 6).

CHAPTER 15 OBSTETRIC AND NEONATAL CONSIDERATIONS

1. Falk, D., Zollikofer, C. & Ponce de León, M. (2012). *Proceedings of the National Academy of Sciences (PNAS).*
2. Gruss. L.T. & Schmitt, D. (2014). The evolution of the human pelvis: changing adaptations to bipedalism, obstetrics and thermoregulation. *Philosophical Transactions of the Royal Society B: Biological Sciences,* 370(1663).
3. Mansfield, F. (2015). Birth and babies. Website.

CHAPTER 16 MARINE ADAPTATIONS IN THE HUMAN KIDNEY

1. Maslin, M.A., Brierley, C.M., Milner, A.M. et al. (2014). East African climate pulses and early human evolution. *Quaternary Science Reviews,* 101: 1–17.
2. Wood, B. (2014). Fifty years after *Homo habilis. Nature,* 508: 501–503.
3. White, T.D., Asfaw, B., Beyene, Y. et al. (2009). *Ardepithecus ramidus* and the paleobiology of early hominids. *Science* 64: 75–86.
4. Cerling, T.E. (2014). *Stable Isotope Evidence for Hominin Environments in Africa. Treatise on Geochemistry,* 2nd ed., pp. 1213–1214.
5. Green, D.J., Gordon, A.D. & Richmond, B.G. (1997). Limb-size proportions in *Australopithecus afarensis* and *Australopithecus africanus. Journal of Human Evolution* 52: 187–200.
6. Williams, M.F. (2011). Marine adaptations in the human kidneys. In: Vaneechoutte, M., Kuliukas, A. & Verhaegen, M. (Eds.). *Was Man More Aquatic in the Past?* Bentham Books, Danvers, MA.

CHAPTER 17 SCARS OF EVOLUTION

1. Pickett, R. (2018). *The Evolution of Mammals.*http://www.bobpickett.org.
2. Crawford, M.A. & Sinclair, A.J. (1972). Nutritional influences in the evolution of the human brain. In: Elliot, K. & Knight, J. (Eds.). *Lipids, Malnutrition and the Developing Brain.* Elsevier, Amsterdam, the Netherlands, pp. 267–292.
3. Morgan, E. (1990). *The Scars of Evolution.* Souvenir Press, London, UK.
4. Sarich, V. & Wilson, A. (1967). Immunological time scale for hominid evolution. *Science,* 158: 1200–1203.
5. Darwin, C. (1871). *The Descent of Man.* John Murray, London, UK.
6. Hardy, A.C. (1960). Was man more aquatic in the past? *New Sci.,* 7: 642–645.
7. Gislen, A., Dacke, M., Kroger, R., Abrahamsson, M. et al. (2003). Superior underwater vision in a human population of sea gypsies. *Current Biololgy,* 13: 833–836.
8. Grey, G. (1841). *Expeditions in Western Australia 1837–1839.* T & W Boone, London, UK.
9. Bailey, G. & Parkington, J. (1988). *The Archaeology of Prehistoric Coastlines.* Cambridge University Press, Cambridge.
10. Herkowitz, H. (2004). *The Lumbar Spine. International Society for Study of the Lumbar Spine.* Ch. 1, p. 5.
11. Corner, M. (2016). Sickness absence in the labour market. *Office for National Statistics.*
12. O'Brien, H., Bourke, J. (2016). Vascular 'safety net' doesn't protect the brains of giraffes from dangerous pressure changes. *Anatomy to you.* January 20.

13. Dinicolantonio, J. (2017). *The Salt Fix.* Piatkus, London, UK.
14. Denton, D.A. (1965). Evolutionary aspects of the emergence of aldosterone secretion and salt appetite. *Physiological Reviews,* 45:245–295.
15. Ng. M., Andrew, T., Spector, T., Jeffery, S. (2005). Linkage to the FOXC2 region of chromosome 16 for varicose veins in otherwise healthy, unselected sibling pairs. *Journal of Medical Genetics,* 42(3): 235–239.

CHAPTER 18 WE ARE WHAT WE EAT

1. Bramble, D. & Lieberman, D. (2004). Endurance running and the evolution of *Homo. Nature,* 432: 345–352.
2. Lieberman, D., Bramble, D. et al. (2007). The evolution of endurance running and the tyranny of ethnography. A reply to Pickering and Bunn. *Journal of Human Evolution,* 53: 439–442.
3. Bender, R., Tobias, P.V. & Bender, N. (2012). The savannah hypotheses: Origin, reception and impact on palaeoanthropology. *History and Philsophy of the Life Sciences,* 34: 147–184.
4. Munro, S. (2013). Endurance running versus underwater foraging: An anatomical and palaeo-ecological perspective. *Human Evolution,* 28: 201–212.
5. Pickering, T.R. & Bunn, H.T. (2007). The endurance-running hypothesis and hunting and scavenging in savanna-woodlands. *Journal of Human Evolution,* 53(4): 434–438.
6. Speth, J.D. (1987). Early hominid subsistence strategies in seasonal habitats. *Journal of Archaeological Science,* 14: 13–29.
7. Dart, R. (1925). *Australopithecus africanus:* The Man-Ape of South Africa. *Nature,* 115: 195–199.
8. Potts, R. (1998). Environmental hypotheses of hominin evolution. *American Journal of Physical Anthropology,* 41: 93–136.
9. Verhaegen, M., Munro, S., Vaneechoutte, M. et al. (2007). The original econiche of the genus *Homo*: Open plain or waterside? In: Munoz, S. (Ed.). *Ecology Research Progress.* Nova, New York, pp. 155–186.
10. Broadhurst, C.L., Crawford, M.A. & Munro, S. (2011). Littoral man and Waterside woman: The crucial role of marine and lacustrine foods and environmental resources in the origin, migration and dominance of *Homo sapiens.* In: Vaneechoutte, M., Kuliukas, A. and Verhaegen, M. (Eds.). *Was Man More Aquatic in the Past?* Bentham Science Publishers, Brussels.
11. Larick, R., Ciochon, R. (1996). The African emergence and early Asian dispersals of the genus *Homo. American Scientist,* 84: 538–552.
12. Arribas, A. & Palmqvist, P. (1999). On the ecological connection between sabre-tooths and hominids: Faunal dispersal events in the Lower Pleistocene and a review of the evidence for the first human arrival in Europe. *Journal of Archaeological Science,* 26: 571–585.
13. Anton, S.C., Leonard, W.R. & Robertson, M. (2002). An eco-morphological model of the initial hominid dispersal from Africa. *Journal of Human Evolution,* 43: 773–785.

14. Kennedy, G.E. (1985). Bone thickness in *Homo erectus*. *Journal of Human Evolution*, 14: 699–708.
15. Cenacchi, T., Bertoldin, T. Et al. (1993). Cognitive decline in the elderly. A double-blind placebo-controlled multicenter study on efficacy of phosphatidyl serine administration. *Aging (Milano)*, 5: 13–33.
16. Stevens, L.J. Zentall, S.S. et al. (1996). Omega-3 fatty acids in boys with behaviour, learning and health problems. *Physiology & Behavior* 59: 915–920.
17. Stordy, B.T. (1997). Dyslexia, attention deficit hyperactivity disorder – do fatty acids help? *Dyslexia Review*, 9: 1–3.
18. Frasure-Smith, N., Lesperance, F. & Julien, P. (2004). Major depression is associated with lower omega-3 fatty acid levels in patients with recent acute coronary syndromes. *Biological Psychiatry*, 55: 891–896.
19. Peet, M.I. & Stokes, C. (2005). Omega-3 fatty acids in the treatment of psychiatric disorders. *Drugs*, 65: 1051–1059.
20. Parker, G., Gibson, N.A. et al. (2006). Omega-3 fatty acids and mood disorders. *American Journal of Psychiatry* 163: 969–978.
21. Nurk, E., Drevon, C.A., Refsum, H. et al. (2007). Cognitive performance among the elderly and dietary fish intake: the Hordaland Health Study. *American Journal of Clinical Nutrition* 86: 1470–1478.
22. Hibbeln, J.R. (2002). Seafood consumption, the DNA content of mother's milk and prevalence rates of postpartum depression: A cross-national, ecological analysis. *Journal of Affective Disorders* 69: 15–29.
23. Catalan, J., Moriguchi, T. et al. (2002). Cognitive deficits in docosahexaeonic acid-deficient rats. *Behavioral Neuroscience*, 116: 1022–1031.
24. Lim, S.Y., Hoshiba, J., Moriguchi, T. & Salem, N. (2005). N-3 fatty acid deficiency induced by a modified artificial rearing method leads to poorer performance in spatial learning tasks. *Paediatric Research* 58: 741–748.
25. Fedorova, I., Hussein, N. et al. An N-3 fatty acid deficient diet affects mouse spatial learning in the Barnes circular maze. *Prostoglandins Leukotrienes Essential Fatty Acids*, 77: 269–277.
26. Cunnane, S.C. & Stewart, K.M. (2010). *Human Brain Evolution. The Influence of Freshwater and Marine Resources*. John Wiley & Sons, Hoboken, NJ.
27. Stewart, K.M. (1994). Early hominid utilisation of fish resources and implications for seasonality and behaviour. *Journal of Human Evolution* 27: 229–245.
28. Clark, J., Beyene, Y. et al. (2003). Stratigraphic, chronological and behavioural contexts of Pleistocene *Homo sapiens*from Middle Awash, Ethiopia. *Nature*, 423: 747–752.
29. Feibel, C., Harris, J. & Brown, F. (1991). Palaeoenvironmental context for the late Neogene of the Turkana Basin. In: Harris, J. (Ed.) *Koobi Fora Research Project Vol. 3. The Fossil Ungulates: Geology, Fossil Artidodactyls, and Palaeoenvironments*. Clarendon Press, Oxford, UK, pp. 321–370.
30. Munro, S. (2010). *Molluscs as Ecological Indicators in Palaeoanthropological Contexts*. PhD Thesis, Australian National University, Canberra, Australia.
31. Yeshurun, R., Bar-Oz, G. & Weinstein-Evron, M. (2007). Modern hunting behaviour in the early Middle Paleolithic: Faunal remains from Misliya Cave, Mount Carmel, Israel. *Journal of Human Evolution*, 53(6): 656–677.
32. Barker, G. (2009). *The Agricultural Revolution in Prehistory: Why Did Foragers Become Farmers?* Oxford University Press, Oxford.

33. Bocquet-Appel, J-P. (2011). When the world's population took off: the spring-board of the Neolithic demographic transition. *Science,* 333: 560–561.
34. Pollard, E., Rosenberg, C. & Tigor, R. (2015). *Worlds Together, Worlds Apart Concise Edition.* Vol. 1. WW Norton, New York.
35. Lewin, R. (2009) The origin of agriculture and the first villagers. In: *Human Evolution: An Illustrated Introduction.* John Wiley & Sons. Malden, MA, p. 250.
36. Armelagos, G. (2014). Brain evolution: The determinants of food choice and the omnivore's dilemma. *Clinical Reviews in Food Science and Nutrition,* 54(10): 1330–1341.
37. Borlaug, N. (1970) The Green Revolution, Peace, and Humanity–Nobel Lecture, December 11, 1970.
38. Hazell, P.B.R. (2009). *The Asian Green Revolution.* IFPRI Discussion Paper. International Food Policy Research Institute.
39. Global Report on Food Crises. (2018). *World Food Programme.*

CHAPTER 19 AN INCREDIBLE JOURNEY

1. Maclarnon, A. (2011). The anatomical and physiological basis of human speech production: Adaptations and exaptations. In: Tallerman, M. & Gibson, K. (Eds.). *The Oxford Handbook of Language Evolution.* Oxford University Press, Oxford.
2. Lieberman, D. (2008). Speculations about the selective basis for modern human craniofacial form. *Evolutionary Anthropology,* 17: 55–68.
3. Deutscher, G. (2006). *The Unfolding of Language: An Evolutionary Tour of Mankind's Greatest Invention.* Holt, New York.
4. Borlaug, N. (1970) The Green Revolution, Peace, and Humanity—Nobel Lecture, December 11, 1970.
5. Hazell, P.B.R. (2009). *The Asian Green Revolution.* IFPRI Discussion Paper. International Food Policy Research Institute.
6. Lozano, R. et al. (2012). Global and regional mortality from 235 causes of death for 20 age groups in 1990 and 2010: A systematic analysis for the Global Burden of Disease Study 2010. *Lancet,* 380(9859): 2095.
7. Andlin-Sobocki, P., Jonsson, J. et al. (2005). Cost of disorders of the brain in Europe. *European Journal of Neurology,* 12(1): 1–27.
8. Crawford, M.A. & Crawford, S.M. (1972). *What We Eat Today.* Neville Spearman, London, UK.
9. Crawford, M.A., Hussein, I., Nyuar, K.B. & Broadhurst, C.L. (2014). The global crisis in brain nutrition and the rise in mental ill-health. *Human Evolution,* 29(1–3): 207–227.
10. Hallahan, B., Hibbeln, J.R. et al. (2007). Omega-3 fatty acid supplementation in patients with recurrent self-harm, single-centre, double-blind randomized controlled trial. *British Journal of Psychiatry,* 190: 118–122.
11. Meyer, B.J., Grenyer, B.F., Crowe, T. et al. (2013). Improvement of major depression is associated with increased erythrocyte DHA. *Lipids.* Jun 4 online.

12. Stuart, K., Soulsby, E.J.L. (2011). Reducing global health inequalities. Part 2: Myriad challenges. *Journal of the Royal Society of Medicine,* 104: 442–448.
13. Fagan, J.J. & Jacobs, M. (2009). Survey of ENT services in Africa: need for a comprehensive intervention. *Global Health Action,* 2: 1932. doi:10.3402/gha. v2i0.1932.
14. Stefan. D.C. (2015). Cancer care in Africa: An overview of resources. *Journal of Global Oncology* 1(1): 30–36.
15. World Health Organisation.(2000) Obesity: Preventing and managing the global epidemic. WHO Technical Support Series. No 894. WHO, Geneva, Switzerland.
16. Eberwine, D. (2002). Globesity: The crisis of growing proportions. *Perspectives of Health,* 7: 3.
17. Prospective Studies Collaboration, Whitelock, G., Lewington, S. et al. (2009). Body-mass index and cause-specific mortality in 900 000 adults: Collaborative analyses of 57 prospective studies. *Lancet,* 373: 1083–1096.
18. Berrington de Gonzalez, A., Harge, P. et al. (2010). Body-mass Index and mortality among 1.46 million white adults. *New England Journal of Medicine* 363: 2211–2219.

Index

Note: Page numbers in italic refer to figures, respectively.

A

abnormal spindle-like microcephaly (ASPM), 132
acne, 64
acquired immune deficiency syndrome (AIDS), 47–48
actinic keratosis, 79
adipose tissue, 67–68
aero-digestive tract, 83, 86, 88
aetiological factor, 105
Afar region, 8, 28
African mammalian evolution, 9
African Negro, 93
agricultural origins and Neolithic Revolution, 181–183
alpha-linoleic acid (ALA), 119
anatomically modern humans (AMHs), 40, 45, 74, 157
anterior ethmoidal nerve, 97
anterior fontanelles, 149
anxiety, 178
ape-human hominin split, 68
Ape/Man transition, *5*
apocrine glands, 56, 63–64
appearance, skin, 62
aquatic adaptations of human babies, 153–154
aquatic ape hypothesis (AAH), 14, 17
The Aquatic Ape Hypothesis (book), 13
aquatic apes, 8
aquatic ape theory (AAT), 13–14, 17
aquatic hypothesis, 73
aquatic mammals, 96
aquatic skin adaptations, 68
arachidonic acid, 59, 178
arboreal apes, 6–7, 86, 89
arboreal habitat, 4
arbor vitae, 117
archaeological fossils, *42*
Ardepithecines, 36
Ardepithecus, 157

Ardipithecus ramidus, 36, 68–69, 157
Ardrey, R., 7
aridification, 28
arrector (or erector) pili muscles, 62–63
artificial intelligence (AI), 145
Attenborough, D., Sir, 13, 16–18, 21
attention-deficit/hyperactivity disorder (ADHD), 178
aural exostoses, 75
Aurignacian creativity, 46–47
Australopithecines, 3, 20, 54, 77, 97, 148, 150, 154, 165, 176
   brains, 115
   family, *36*
   fossils, 30
Australopithecus ('southern ape'), 6, 9, 36–39, 69, 157
Australopithecus afarensis, 36, 157
Australopithecus africanus, 39, 150, 162
   skull, *149*
autonomic nervous system, 62, 65–66

B

Barr, T., 111
Bartholinus, T., 92
basal cell carcinoma (BCC), 79
Belgraver, P., 110
bipedal ape, 9
bipedal heritage, 172–173
bipedalism, 5–6, 52
   adaptations to, 163–165
bipolar disorder, 178
The Birth of Homo, the Marine Chimpanzee (book), 11
Black Hole, *4*, *5*
blood pressure and salt regulation, 168–170
body louse, 77
bonobo, 5, 51, 68, 149–150, 153, 186–187
bony ear swellings, 103, 105
bradycardia, 97

brain, human, 115–116, *116*
  areas, *118*
  evolution, genetic factors, 131–133
  lipids, 118–121
  size and perinatal considerations, 150–151
  speech and language areas, *144*
  structure, 116–118
Bronowski, J., 51
brown fat, 54

C

carcinogenesis, 79
carotid, *125*
  rete, 168
cellular control system, 27
cellular inflammatory process, 107–108
cerebellum, 117, *117*
cerebral blood flow, *124*
cerebrum, 117
cervical muscles, 96
cervico-occipital articulation, 96
cetacean family, 9
cetaceans, 96
Changizi, M., 66
Cheddar Gorge Man, 78
chimpanzees, 3, 68, 78
  epiglottis and soft palate, *138*
  swimming and diving, 81
chronic obstructive airway disease
       (COAD), 166
climatic adaptation, 56
Cloquet, H., 92
coastal migration and worldwide dispersal,
       127–128
coccyx, 166
cognitive revolution, 4, 128–131, 179–181
The Complete World of Human Evolution
       (book), 11
computerized tomography (CT), 21, 105
corticosterone, 169
crank anthropology, 17
creationism, 17
Crick, F., 26
Cro-Magnons, 44–45, *45*
Crural Index, 44–45
Cushing's disease, 169

D

Dart, R., 6–7, 150, 162
Darwin, C., 1, 24, 59, 69, 123

Darwin's theory of evolution of humans and
       apes, 88
De Humani Corporis Fabrica (book), 85
Denisovians, 32
  evolution and spread, *42*
deoxy-ribonucleic acid (DNA)
  analysis, 78
  hybridization, 37
  three-dimensional structure, 26
depression, 178
Derbyshire neck, 120–121
Der Eigenweg des Menschen (*The Unique
       Road to Man*), 7
dermis, 67
The Descent of Man (book), 2, 24, 32, 123, 163
The Descent of Woman (book), 13, 76
diverticulosis, 140
diving reflex, 96–98
docosahexaenoic acid (DHA), 59, 119, 162, 178
domestication, 126–127
dominant species, 29
dromedary camel, *160*
Dudley-Buxton, L.H., 100
dyslexia, 178

E

ear, nose and throat (ENT) surgeons, 103
ear canal, *104*, *106*
  bone abnormalities, 105
  embryology, 105–108
  exostoses in modern populations, 110–111
eardrum, 106
ear exostoses, 103
  archaeological populations, 111
  vital fossil 'missing link,' 112–113
early bipedal hominins, 35–36
early hominin, *91*
East African Great Rift Valley, 155, *156*
eccrine
  glands, 57, 64–65
  sweat-cooling system, 58
  sweat glands, 57
  sweating, 57
echolocation, 108–109, *109*
ecological structure, 28
ecological waterside, 33
elephant seals, 100
endogenous heat, 76
epidermal growth factor receptor (EGFR), 79
epigenetic forces in evolution, 27
epiglottis and soft palate, 138

epithelial mucosa, 100
epithelium, 92
ethmoid sinuses, 90–92
ethmo-turbinals, 90
eumelanin, 77
eumelanin protein, 79
evolution
    bipedal heritage, 172–173
    bipedalism, adaptations to, 163–165
    blood pressure and salt regulation,
        168–170
    epigenetic forces in, 27
    hernias, haemorrhoids and prolapses,
        170–171
    hominin evolution, 162–163
    humans' place in, *36*
    lumbar disc and sciatic problems,
        165–166
    mammalian evolution, 161–162
    varicose veins, 171–172
    vertigo and neck pain, 167–168
evolutionary 'Black Hole,' *4, 5*
Evolution – The Human Story (book), 11
existence, conditions of, 27
exogenous radiation, 76
exostoses, *see* ear exostoses
external auditory canal, 111
    division, 105
    exotoses, 105
external nose, 99–100

F

facial growth theory, 94
Fahlman, A., 141
Falk, D., 149
fatty acid, 119
Fernel, J., 83
fire, 74
fontanelles and skull sutures, 148–150
food crisis, 183
food preservation and cooking, 175
fossil analysis, 24
fossils, 21
Franklin, R., 26
frontal and sphenoidal sinuses, 90
fundamental sound frequency (F0), 137
fur coat, 73–74
    christening ceremony, 75–77
    hominid's 'christening ceremony,' 81–82
    skin colour, ultraviolet irradiation and
        cutaneous cancer, 77–81

G

Galapagos Islands, 24
Galen, 85
Gannon, P., 144
Gee, H., 17–18
genes, 26
genetic DNA sequences, 23
genetic introgression, 46
genetic mutations, 24
genome, 27
genotype, 26
global food crisis, 183
Gorge, C., 82
gorilla, 3, *52*
Gould, S.J., 136
Great Rift Valley, 27, 30–33, *31*, 36, 177
    Australopithecine sites, *32*
    human kidney, 155–158
ground-dwelling mammals, 53

H

Hadar settlement, 20, *42*
haemeostatic controls, 96–97
haemoglobin, 97
haemorrhoids, 171
hair distribution in humans, 76
hairlessness, 52
hairy savages, 43
Haller, A., 92
haplogroup, 129
Harding, R., 77
Hardy, A., Sir, 7, 9, 11, 52, 62
Harvey, W., 171
Hawking, S., 1
Hawks, J., 17
hay fever, 85
head and neck squamous cell carcinoma
        (HNSCC), 79
head louse, 77
hearing mechanism in aquatic and
        semiaquatic mammals, 108–110
Herdlycke, A., 105
heredity, principles of, 25
hernias, haemorrhoids and prolapses, 170–171
heterozygous, 26
hippocampus, 145
hirsute males, 76
hobbit, 41
Holocene epoch, 49
hominid, 141–142

hominins, 84
   evolution, 162–163
   speciation, 158
hominoid fossils, 37
Homo denisova, 41–42
Homo erectus, 9–10, 39–41, 73–74, 150, 177
   cranium, 40
Homo ergaster, 39, 41, 69
Homo family, *36*
Homo floresiensis, 32, 41
Homo fossils, 111
Homo habilis, 39, 69, 115, 150
Homo heidelbergensis, 40–41, 69
Homo neanderthalensis, 41, 43–49, *44*,
     45, 115
Homo rhodesiensis, 40
Homo sapiens, 3, 9, 22, 39, 43, 46, *52*, 73
   emergence, 4
   in Indonesia, 41
Homo stupidus, 43
homozygous, 26
hooded seals, 110
Hooker, J., 1
human
   external ear canal, embryology, 105–108
   genome, 23
   keratinocytes, 79
   neocortex, 124
   scalp skin, *67*
human brain, 115–116
   areas, *118*
   evolution, genetic factors, 131–133
   lipids, 118–121
   structure, 116–118
human evolution, theories, 1–3
   humanity's place, 3–6
   humankind and climatic change, 9–11
   savannah theory, 6–7
   waterside theory, 7–9
human kidney, *159*
   Great Rift Valley, 155–157
   waterside aquatic habitat, 158–160
humankind and climatic change, 9–11
human papilloma virus (HPV), 80
human populations in the Pleistocene era,
    38–39
   Homo denisova, 41–42
   Homo erectus, 39–40
   Homo floresiensis, 41
   Homo habilis, 39
   Homo heidelbergensis, 40–41
human skull buoyancy, 95–96

diving reflex, 96–98
external nose, 99–100
   nasal valve, 98
   nitric oxide, 96–98
   nose and sinuses in waterside habitat,
     100–101
   paranasal sinuses, 96–98
Huxley, T., 1
hybridization, 37
hydrocortisone (cortisol), 169
hyoid bone, 139
hyperadrenocorticism, 170
hypertension, 125
hypodermis, 66
hypoxia, 98
   in fishes, 97

I

Ice Ages, 5
immune-compromised patients, 79
immunological protection, skin, 62
injury repair, skin, 62
intelligence, 73–74
inter-canthal distance, 90
inter-vertebral discs, 166
inter-vertebral foramina, 167
introgression, 46
ischaemia, 97
isotope, 18

J

Johanson, D., 16, 36, 162

K

Kennedy, G.E., 110
Koobi Fora, 40
Korner, O., 110
Kostyuk, V., 79
Kroon, D.F., 111

L

labyrinth, 87, 90–91, 100
Laetoli footprints, *38*
land-based primates, 58
Langdon, J., 14
lanugo, 151
   and fat babies, 151–152
large adenoids, 85

laryngeal/tongue-base complex, 141
larynx, 88
late Miocene drought, *28*, 28–30
Leakey, M., 37, 162
Leakey, R., 30, 39
Lebanon's Beqaa Valley, 30
Leonardo da Vinci's *Two views of the
        skull*, *86*
Levant, 179, *180*
Levin, B., 14, *15*
Lewis, G.E., 4
lice, 77
Lieberman, P., 138
lightening skull, 93–94
lipoproteins, 121
littoral (waterside) habitat, 52
Llardo, M., 97
long-chain polyunsaturated fatty acid
        (LC-PUFA) deficiencies, 178
'Lucy' (*Australopithecus afarensis*), 36, *37*,
        38
lumbar disc and sciatic problems, 165–166
Lundberg, J.O., 98
Lyell, C., 38

M

magnetic resonance imaging (MRI), 118
mammalian evolution, 161–162
mammalian fat, 54
manual dexterity, 74
marine and lacustrine foods, 177–178
        agricultural origins and Neolithic
                Revolution, 181–183
        cognitive revolution, 179–181
marine mammals, 108
Maslin, M., 17
mastication process, 140
mastoid process, 140
maxillary sinus, 84, 89
maxillary sinusitis, 87
Mbuti Pygmies, 56
MC1R gene, 77, 82
MC1R genetic mutation, 79
Mendel, G., 25
        plant experiments, 25–27, *26*
mental ill health, 189–190
metopic suture (MS), 150
microcephalin, 132
middle ear infections, 85
milieu extérieur du corps, 10
Miocene drought, 28–30

mitochondria, 129
modern ape, *91*
modern human, *91*
Montagna, W., 58
Morgan, E., 3, 13, 15, 165
morphological mechanisms, 9–10
Morris, D., 13, 75
Mousterian technology, 47
Mozambique, 30
mucus secretion theory, 92
Munro, S., 111
mutation, 27
myoglobin, 97

N

The Naked Ape (book), 13, 75
naked apes, 61
        apocrine glands, 63–64
        eccrine glands, 64–65
        skin, evolutionary evidence, 68–71
        skin, structure and function of, 61–63
        subcutaneous fat, 67–68
        wrinkly fingers, 65–67
nasal valve, 98
National Health Service (NHS), 80
Nature Communications (journal), 145
Neanderthals, 43–49, *44*, *45*, 89, 150
negative buoyancy, 96
Negus, V., Sir, 86, 95
neocortex, 117, 121
neurogenesis, 119
neuronal migration, 119
New Scientist (magazine), 7
Nilotic people of Africa, 56
nitric oxide (NO), 96–98
nonhuman primates, 64
non-primate species, 57
nose and sinuses in waterside habitat,
        100–101

O

obstetric and neonatal considerations, 147
        aquatic adaptations of human babies,
                153–154
        brain size and perinatal considerations,
                150–151
        fontanelles and skull sutures, 148–150
        lanugo and fat babies, 151–152
        obstetric dilemma, 148
        vernix caseosa, 152–153

obstetric dilemma, 148
Odent, M., 11
Ohala, J., 139
Olduvai Gorge, 5, 20–21, 30–31, 39, 158
olfactory theory, 92
open anterior-facing nasal aperture, 99–100
The Origin of Species (book), 1
*The Origin of Species by Means of Natural
       Selection* (book), 24
Orronin, 157
Orrorin tugenensis, 35
osteomata, 11
otolaryngologists, 110
'Out of Africa' theory, 45

P

pachyosteosclerosis, 18, 111, 177
Pacinian corpuscles, 66
palaeoanthropology, 14, 17, 103
palaeolithic tool from Mount Carmel, *181*
paleo-mammalian brain, 116
pale-skinned hominin ancestors, 80
Pan troglodytes, *52*
papyrus, 18
paranasal sinuses, 84, *84*, 96–98
       drainage from, *87*
       ethmoid sinuses, 90–92
       evolution and comparative anatomy, 89
       frontal and sphenoidal sinuses, 90
       maxillary sinus, 89
       theories of, 92–94
Paranthopus, 157
PDE10A, 97–98
periosteum, 107
phenotype, 25–26
pheomelanin, 77
photo-protection, 79
physiological development, 162, 187
piles, 171
planum temporale, 144
Pleistocene, 38
Pleistocene era, 38–39, 111
Pleistocene Ice Age, 56
pliocene, *4*
       epoch, 4, 29
Plio-Pleistocene period, 10
pollination, 25
posterior fontanelles, 149
post hoc ergo propter hoc, 6
Potts, R., 9
pre-hominid ape, 3, *91*

primate quadruped forest-based ancestors, 9
primates, 57
primitive hindbrain, 124
proboscis monkey, 8, 10, *8*
Proetz, A.W., 92–94
prolapse, 170
prosimians, 64
protective barrier, skin, 61
protective hair covering, 73–74
proto-reptilian brain, 116
pseudoscience, 17
pubic louse, 77

R

radiation protection, skin, 62
Ramapithecus, *4*, 89
'recessive' green factor, 25
resonance theory, 92
respiratory adaptations, 68
respiratory turbinates, 100
rhinitis, 85
Rightmire, G.P., 111
Roberts, A., 11, 17
Roche, A.F., 105
Ruxton, G.D., 76

S

Sahelanthropus, 157
Sahelanthropus tchadensis, 35
salt (sodium chloride), 158
salt-excreting eccrine sweat glands, 160
Sanger, F., 26
Savannah Ape, 4
Savannah Hypothesis (SH), 15–16
savannah theory, 6–7, 14, 95
       inconsistencies in, 13
       weaknesses of, 7
savannah theory of evolution, 86
scent-producing function, 56
Schagatay, E., 141
schizophrenia, 178
sciatic pain, 166
sea lions, 70
seals, 70
       nostrils, 98
       skin, *70*
sebaceous glands, 64
sebum, 64
sedges, 18
semiaquatic habitat, 20, 108

semiaquatic hominins, 66
semiaquatic lifestyle, 81
semiaquatic mammals, 48, 63, 76, 109, 151, 153
senile dementia, 178
sensation, skin, 62
sensory nerves of face, 97
Seymour, R., 97, 123
sheep kidney, *159*
Shunkov, M., 41
sinuses, 84, *84*, 90–92, 96–98, 100–101
    drainage from, *87*
    ethmoid sinuses, 90–92
    evolution and comparative anatomy, 89
    frontal and sphenoidal sinuses, 90
    maxillary sinus, 89
    theories of, 92–94
Skillern, R.H., 95
skin
    colour, 77
    evolutionary evidence, 68–71
    structure and function of, 61–63, *63*
    thermoregulation, 63, 65
skin squamous cell carcinoma (SCC), 79
skin surface lipid (SSL) mediator, 79
skulls comparison, *85*
social inequality and poverty, 190–191
Sokolov, V., 54, 58, 69, 81
speech and language, 187–189
Spencer, H., 24
squalene, 79
sterno-mastoid muscle, 140
Stringer, C., 11, 45
subcutaneous fat, 67–68, 75, 96, 152
subtropical savannah, 68
sudden infant death syndrome (SIDS), 85, 138
sunstroke, 57
supraglottic vocal tract, 137
surfer's ear, 103–104
    ear canal bone abnormalities, 105
    ear exostoses as vital fossil 'missing link,'
        112–113
    external ear canal exostoses in modern
        populations, 110–111
    external ear exostoses in archaeological
        populations, 111
    hearing mechanism in aquatic and
        semi-aquatic mammals, 108–110
    human external ear canal, embryology,
        105–108
sweat glands, 65
sweat secretion, 57
sympathetic nervous system, 57

T

tactile sensibility, 65
tailless amphibians, 98
tectonic activity, 156–157
temperature of earth, *29*
temperature regulation, skin, 62
temporomandibular lesions, *112*
terra firma, 6
thermal insulation theory, 92–93
thermal sweating in humans, 65
thermoregulation, 55–58, 79
    system, 62
Thompson, A., 100
Tobias, P., 14–17, 52
tonsillitis, 85
toothed whale, 108–109, *109*
trigeminal cardiac reflex (TCR), 97
turnover pulse hypothesis, 9

U

ungulate, 89, 94
unique human modifications, 147
upper aero-digestive tract, 85
Urba, E., 136
UV-sensitive receptor, 79

V

Van Gilse, P.H.G., 110
variability selection hypothesis, 9
varicose veins, 171–172, *172*
vasoconstriction, 97
Venus figurines, 54, *55*
Verhaegen, M., 15, 111
vernix caseosa, 152–153, *153*
vertebral arteries, *125*
    canals, 125
vertigo and neck pain, 167–168
vocal communication, 136
vocal cords, 136
voice, human larynx and evolution of,
        135–137
    changes in brain for speech and language,
        143–146
    delayed descent, 143
    diving and breath-holding in hominids,
        141–142
    upper airway and digestive tracts,
        137–141
Vrba, E.S., 9

W

Wallace, A., 1
warm-blooded creatures, 53
water, proximity to, 16
waterside ape, 51–52
    big brains, 58–59
    bipedalism, 53
    nakedness, 53–54
    subcutaneous fat, 54–55
    thermoregulation, 55–58
The Waterside Ape, 17–18
waterside/aquatic ape hypothesis, 11
waterside aquatic habitat, human kidney and,
        158–160

waterside habitat, 81
waterside theory, 7–9, 20, 76
Watson, J., 26
Westenhofer, M., 7
Wheeler, P., 76
white fat, 54
Wilkins, M., 26
Wilkinson, D.M., 76
Wood-Jones, F., 89
wrinkling of skin, 66
wrinkly fingers, 65–67

Z

Zhoukoudian, 111

Printed and bound by CPI Group (UK) Ltd, Croydon, CR0 4YY

23/10/2024

01778263-0003